General Relativity

I. B. Khriplovich

General Relativity

 Springer

I. B. Khriplovich
Budker Institute of Nuclear Physics
Novosibirsk, Russia 630090
I.B. Khriplovich@inp.nsk.su

ISBN-13: 978-1-4419-2065-2   e-ISBN-13: 978-0-387-27406-5
e-ISBN-10: 0-387-27406-5

Printed on acid-free paper.

Printed in the United States of America.   (EB)

9 8 7 6 5 4 3 2 1

springeronline.com

Preface

The book is based on the course on general relativity given regularly at the Physics Department of Novosibirsk University. The course, lasting for one semester, consists of 32 hours of lectures and 32 hours of tutorials, plus homework of 10 – 12 problems. The exam is passed by 30 – 35 students. The results of the homework and exam give good reasons to believe that at least 20 – 25 of these students really digest the subject.

The course requires of students the knowledge of analytical mechanics and classical electrodynamics, including special relativity. Only chapters 7 and 10 of the book are in this respect exceptions: the acquaintance with the notion of spin is useful for studying chapter 7, the fundamentals of thermodynamics and quantum mechanics are necessary for the last chapter. But these parts of the book can be skipped without any loss for understanding all other chapters.

The book (as well as the course itself) is influenced essentially by the monograph by L.D. Landau and E.M. Lifshitz, *The Classical Theory of Fields*, (Butterworth – Heinemann, 1975). However, I strived to make the exposition as close as possible to a common university course of physics, to make it accessible not only for theorists.

The book is also influenced by the course of lectures by A.V. Berkov and I.Yu. Kobzarev, *The Einstein Theory of Gravity*, (Moscow, MEPhI, 1989, in Russian). In particular, I borrowed from it the derivation of the equations of motion from the Einstein equations (going back to P.A. Dirac and L.D. Landau), the derivation of the Schwarzschild solution (belonging to H. Weyl), as well as the discussion of cosmology.

However, the book contains a lot of material absent in the above sources. Of course, the selection of these topics was determined to a large extent by my own scientific interests. Among these subjects are the gravitational lensing, the signal retardation in the gravitational field of the Sun, the Reissner – Nordström solution, some spin effects, the resonance transformation of an electromagnetic wave into a gravitational one, the gravitational radiation of ultrarelativistic particles, the entropy and temperature of black holes.

The book contains many problems.

In fact, a considerable part of the content of the book was not presented at the lectures, but was discussed at the tutorials. Moreover, in some cases the succession of presentation is dictated by the necessity to create in good time a necessary basis for tutorials and homework.

It is worth mentioning also that some questions considered in the book are sufficiently difficult, though they require no extra knowledge. Usually these questions are discussed neither at the lectures nor at the tutorials. There are also difficult problems which are not obligatory. All this material is intended for an independent work of those students who are most seriously interested in the subject.

One cannot overestimate the imprint made on the book by the collaboration with A.I. Chernykh and V.M. Khatsymovsky in teaching general relativity, this collaboration lasted for many years. In particular, some problems in the book belong to them. A.I. Chernykh, V.M. Khatsymovsky, and V.V. Sokolov also made many useful comments on the manuscript.

The lively interest of numerous students was extremely important for me.

Some original results presented in the book were obtained in collaboration with R.V. Korkin, A.A. Pomeransky, E.V. Shuryak, O.P. Sushkov, and O.L. Zhizhimov.

To all of them I owe my deep and sincere gratitude.

In the fall of 2003, I lectured on general relativity at Scuola Normale Superiore, Pisa, Italy. The major part of translating the book into English was done during this visit. I recall with gratitude the warm hospitality extended to me at Scuola Normale and the lively interest of its students to my lectures.

Novosibirsk,   *Iosif Khriplovich*
October 2004

Contents

1 Introduction 1

2 Particle in Gravitational Field 5
   2.1 Electrodynamics and Gravitation 5
   Problem 6
   2.2 Principle of Equivalence
        and Geometrization of Gravity 6
   2.3 Equations of Motion of Point-Like Particle 7
   2.4 The Newton Approximation 9

3 Fundamentals of Riemann Geometry 11
   3.1 Contravariant and Covariant Tensors. Tetrads 11
   Problems 13
   3.2 Covariant Differentiation 13
   Problems 15
   3.3 Again Christoffel Symbols and Metric Tensor 15
   Problems 17
   3.4 Simple Illustration of Some Properties
        of Riemann Space 18
   3.5 Tensor of Curvature 19
   Problems 21
   3.6 Properties of the Riemann Tensor 21
   Problems 23
   3.7 Relative Acceleration of Two Particles
        Moving Along Close Geodesics 24

4 Einstein Equations 27
   4.1 General Form of Equations 27
   4.2 Linear Approximation 28
   4.3 Again Electrodynamics and Gravity 29
   4.4 Are Alternative Theories of Gravity Viable? 31

5 Weak Field. Observable Effects 33
 5.1 Shift of Light Frequency
       in Constant Gravitational Field 33
 5.2 Light Deflection by the Sun 34
 5.3 Gravitational Lenses 35
 Problem 38
 5.4 Microlenses 38
 Problem 39

6 Variational Principle. Exact Solutions 41
 6.1 Action for Gravitational Field.
       Energy-Momentum Tensor of Matter 41
 Problems 44
 6.2 Gravitational Field of Point-Like Mass 44
 Problems 46
 6.3 Harmonic and Isotropic Coordinates.
       Relativistic Correction to the Newton Law 46
 Problems 48
 6.4 Precession of Orbits in the Schwarzschild Field 49
 Problems 51
 6.5 Retardation of Light in the Field of the Sun 51
 Problems 53
 6.6 Motion in Strong Gravitational Field 54
 Problems 56
 6.7 Gravitational Field of Charged Point-Like Mass 57

7 Interaction of Spin with Gravitational Field 61
 7.1 Spin-Orbit Interaction 61
 Problem 62
 7.2 Spin-Spin Interaction 63
 Problems 65
 7.3 Orbit Precession Due to Rotation of Central Body 66
 Problems 67
 7.4 Equations of Motion of Spin in Electromagnetic Field 67
 Problems 71
 7.5 Equations of Motion of Spin in Gravitational Field 71
 Problems 75

8 Gravitational Waves 77
 8.1 Free Gravitational Wave 77
 Problems 80
 8.2 Radiation of Gravitational Waves 80
 Problems 84
 8.3 Gravitational Radiation of Binary Stars 84
 Problems 85

8.4   Resonance Transformation of Electromagnetic Wave
to Gravitational One 86
Problem 87
8.5   Synchrotron Radiation of Ultrarelativistic Particles
Without Special Functions 87
Problem 90
8.6   Radiation of Ultrarelativistic Particles
in Gravitational Field 90
Problems 92

9   General Relativity and Cosmology 93
9.1   Geometry of Isotropic Space 93
Problems 96
9.2   Isotropic Model of the Universe 96
Problems 99
9.3   Isotropic Model and Observations 99
Problem 101

10 Are Black Holes Really Black? 103
10.1 Entropy and Temperature of Black Holes 103
Problem 109
10.2 Entropy, Horizon Area, and Irreducible Mass.
Holographic Bound. Quantization of Black Holes 109
Problems 114

Index 115

1

Introduction

General relativity (GR) is the modern theory of gravity relating it to the curvature of the four-dimensional space-time.

In its so to say classical version, the theory of gravity was created by Newton as early as in the seventeenth century, and up to now it serves mankind faithfully. It is quite sufficient for many, if not most, problems of modern astronomy, astrophysics, and space research. Meanwhile, its inherent flaw of principle was clear already to Newton himself. This is a theory with action at a distance: in it the gravitational action of one body on another is transmitted instantaneously, without any retardation. The Newton gravity is related to general relativity in the same way as the Coulomb law is related to Maxwell electrodynamics. Maxwell has expelled action at a distance from electrodynamics. In gravity it has been done by Einstein.

One should start with the remarkable work by Einstein of 1905 where special relativity was formulated and development of the classical electrodynamics was completed. Certainly this work had its predecessors, one cannot but mention among them Lorentz and Poincaré. Their papers contain many elements of special relativity. However, clear understanding and a complete picture of the physics of high velocities appeared only in the mentioned work by Einstein. This is no accident that up to now, in spite of the existence of many excellent modern textbooks, this work can be recommended for the first acquaintance with the subject even for freshmen.

As to GR, all its fundamentals were created by Einstein.

However, the anticipation that physics is related to the curvature of space can be found in the works by remarkable scientists of nineteenth century: Gauss, Riemann, Helmholtz, Clifford. Gauss came to the ideas of non-Euclidean geometry somewhat earlier than Lobachevsky and Bolyai, but did not publish his investigations in this field. Gauss not only believed that "the geometry should be put in the same row not with arithmetics that exists purely *a priori*, but rather with mechanics". He tried to check experimentally, by means of precision (for his time) measurements, the geometry of our space. The ideas by Gauss inspired Riemann who believed that our space is really curved (and even discrete at small distances). Strict bounds on the space curvature were obtained from astronomic data by Helmholtz. And Clifford thought of matter as a sort of ripples on a curved space.

However, all these brilliant guesses and predictions were evidently premature. Creation of the modern theory of gravity was inconceivable without the special relativity, without deep understanding of the structure of classical electrodynamics, without deep realization of the unity of space-time. As mentioned already, GR was created essentially by the efforts of a single person. The Einstein way to the construction of this theory was long and torturous. While his work of 1905 "On the Electrodynamics of Moving Bodies" had appeared as if immediately in a complete form, leaving out of sight of readers long reflections and heavy work of the author, with GR it was the other way around. Einstein started working on it in 1907, and his way to GR took a few years. It was a way of trial and error that can be traced at least partially through his publications during those years. The problem was finally solved by Einstein in two works presented at the meetings of the Prussian Academy of Sciences in Berlin on 18 and 25 November 1915.

At the last stage of the creation of GR, Hilbert contributed to it by formulating the variational principle for the gravitational field equations. In general, the importance of mathematics and mathematicians for GR is truly great. Its apparatus, the tensor analysis, or the absolute differential calculus, was developed by Ricci and Levi-Civita. Mathematician Grossmann, a friend of Einstein, introduced him to this technique.

However, GR is a physical theory and a completed one. It is completed in the same sense as classical mechanics, classical electrodynamics, and quantum mechanics. Like those theories, it gives unique answers to physically reasonable questions and gives clear predictions for observations and experiments. However, the applicability of GR, as well as that of any other physical theory, is limited. So, beyond its applicability limits the superstrong gravitational fields remain where quantum effects are of importance. Complete quantum theory of gravity does not exist.

GR is a remarkable physical theory because it is based essentially on a single experimental fact, and this fact had been known for a long time, well before the creation of GR (all bodies fall in the gravitational field with the same acceleration). It is remarkable because it was created essentially by a single person. But first of all, GR is remarkable because of its unusual internal harmony and beauty. It is no accident that Landau said: one can recognize in a person a true theoretical physicist by his admiration experienced at the first acquaintance with GR.

Until about the middle of 1960th GR was to a considerable degree beyond the main stream of the development of physics. And the development of GR itself in no way was too active, being confined mainly to clarification of some subtleties and theoretical details, as well as the solution of important, but still sufficiently special problems. I recall a respectable physicist who did not recommend young theorists to work in GR. He said: "This is a science for elderly people".

Possibly one reason is because GR arose in a sense too early, Einstein was ahead of his time. On the other hand, already in his work of 1915 the theory was formulated in a sufficiently complete form. No less important is the fact that the observational base of GR for a long time remained very narrow. Corresponding experiments are extremely difficult. It is sufficient to recall that experimentalists succeeded in measuring the red shift only in a half a century after it had been predicted by Einstein.

However, at present GR is developing rapidly. This is a result of tremendous progress of observational astronomy and development of the experimental technique. On the other hand, researches in quantum gravity are in the forefront of the modern theoretical physics.

I hope that the present volume will serve as a comprehensible introduction to this exciting field of exploration of Nature.

2

Particle in Gravitational Field

2.1 Electrodynamics and Gravitation

We start with the comparison between the equations of motion of a point-like particle in the electromagnetic and gravitational fields. We will compare as well the equations for these fields.

The equation of motion for a particle of a mass m and a charge e in an electromagnetic field $F_{\mu\nu}$ is well known:

$$m\frac{du^\mu}{ds} = eF^{\mu\nu}u_\nu \, . \tag{2.1}$$

Here $u^\mu = dx^\mu/ds$ is the four-velocity of the particle; $ds^2 = \eta_{\mu\nu}dx^\mu dx^\nu$ is the four-dimensional interval in the Minkowski space; the diagonal metric tensor in this space is chosen as $\eta_{\mu\nu} = \mathrm{diag}\,(1, -1, -1, -1)$; the velocity of light is put to unity, $c = 1$.

The equations of the electromagnetic field are

$$\partial_\mu F^{\mu\nu} = 4\pi j^\nu, \quad F_{\mu\nu} = \partial_\mu A_\nu - \partial_\nu A_\mu \, . \tag{2.2}$$

Here A_μ is the electromagnetic vector-potential, and the four-dimensional current density of point-like particles (marked by index a) is

$$j^\nu = \sum_a e_a \, \delta(\mathbf{r} - \mathbf{r}_a(t)) \, u_a^\nu \, \frac{ds}{dt} \, . \tag{2.3}$$

In the Lorentz gauge $\partial_\mu A^\mu = 0$ the Maxwell equation (2.2) reduces to

$$\Box A^\nu = 4\pi j^\nu; \quad \Box = \partial_\mu \partial_\mu = \partial_t^2 - \Delta \, . \tag{2.4}$$

Equations (2.1) and (2.2), taken together with initial conditions for charges and fields on a space-like surface, determine completely the evolution of a system. The equations of electrodynamics are linear, for electromagnetic fields the superposition principle is valid.

The equations of motion of a point-like particle in an external gravitational field are

$$\frac{du^\mu}{ds} = -\Gamma^\mu_{\nu\varkappa} u^\nu u^\varkappa. \tag{2.5}$$

In the case of a weak gravitational field the symbol $\Gamma_{\mu,\nu\varkappa}$ is expressed as follows through its potential, symmetric second-rank tensor $h_{\mu\nu}$:

$$\Gamma_{\mu,\nu\varkappa} = \frac{1}{2}\left(\partial_\nu h_{\mu\varkappa} + \partial_\varkappa h_{\mu\nu} - \partial_\mu h_{\nu\varkappa}\right). \tag{2.6}$$

The equations for a weak gravitational field (in a gauge analogous to the Lorentz one) are

$$\Box h_{\mu\nu} = -16\pi k \left(T_{\mu\nu} - \frac{1}{2}\eta_{\mu\nu}T^\varkappa_\varkappa\right). \tag{2.7}$$

Here

$$k = 6.67390(9) \cdot 10^{-8} \text{ cm}^3 \cdot \text{g}^{-1} \cdot \text{s}^{-2} \tag{2.8}$$

is the Newton gravitational constant, and the energy-momentum tensor of point-like particles is

$$T_{\mu\nu} = \sum_a m_a\, \delta(\mathbf{r} - \mathbf{r}_a(t))\, u^\mu_a u^\nu_a \frac{ds}{dt}. \tag{2.9}$$

The similarity to electrodynamics is evident, however the distinction from the latter is in fact very essential.

The point is that it is the charges that serve as a source of the electromagnetic field. But the electromagnetic field by itself is neutral, it bears no charge. As to the gravitational field, its source is energy, however, the gravitational field possesses energy by itself. Therefore, the gravitational field equations are in fact nonlinear. The linear equations (2.6) and (2.7) are valid, as has been pointed out already, for weak fields only.

Problem

2.1. What is the behavior of the current density (2.3) and the energy-momentum tensor (2.9) under the Lorentz transformations? How does $\delta(\mathbf{r} - \mathbf{r}_a(t))$ transform?

2.2 Principle of Equivalence and Geometrization of Gravity

GR is based on a clear physical principle, on a firmly established experimental fact known already to Galileo: all bodies move in the field of gravity (if the resistance of the medium is absent) with the same acceleration, the trajectories of all bodies with the same velocity are curved alike in the gravitational field. Because of this, in a freely falling elevator no experiment can detect the gravitational field. In other words, in the reference frame freely moving in a gravitational field there is no gravity in a small space-time region. The last statement is one of the formulations of the equivalence principle. This property of the gravitational field is far from being trivial. It is sufficient to recall that for the electromagnetic field the situation is completely different. There exist for instance non charged, neutral bodies that do not feel at all the electromagnetic field. However, there are no gravitationally-neutral bodies, there exist neither rulers nor clocks that would not feel the gravitational field. There are no objects that could be identified in this field with straight lines, as this is the case in the Euclidean geometry. Therefore, it is natural to believe that the geometry of our space is non-Euclidean.

Still, in the local frame connected with a freely falling elevator the metric remains the Minkowski one, and intervals of time and space coordinates are measured by usual clocks and rulers. However, it cannot be done in the whole space-time if the gravitational field is present. The coordinates $x^0 = t$, x^1, x^2, x^3 are just space-time labels. They are continuous, i.e. close values of x^μ correspond to two close points. The general form of the interval is

$$ds^2 = g_{\mu\nu}dx^\mu dx^\nu; \qquad (2.10)$$

here and below the summation over repeated indices is implied. The symmetric second-rank tensor $g_{\mu\nu}(x)$ defines the Riemann space. Since in a locally inertial frame it reduces to $\eta_{\mu\nu} = \mathrm{diag}\,(1, -1, -1, -1)$, the rank of the matrix $g_{\mu\nu}(x)$ is 4, and its signature is (-2).

A reasonable physical realization of a coordinate frame in the Riemann space is collisionless dust. Each dust particle has a space label x^m, $m = 1, 2, 3$, as well as an arbitrary going clock. The coordinates are continuous, and on some space-like surface we put $x^0 = 0$ for all clocks. In such a physically reasonable metric $g_{00} > 0$, the matrix g_{mn} of the metric on the surface $x^0 = 0$ has the rank 3 and the signature (-3).

The metric created by a well-localized distribution of gravitating masses is asymptotically flat. However our Universe as a whole could be non-Euclidean.

2.3 Equations of Motion of Point-Like Particle

In special relativity the trajectory of a free point-like particle, moving between two points A and B, is determined by the variational principle

$$\delta \int_A^B ds = 0, \qquad (2.11)$$

where ds is the interval in the Minkowski space. Since the action of a gravitational field reduces to a change of the space-time metric, in this field the variational principle has the same form (2.11), but now ds is the interval in the Riemann space and is defined by formula (2.10). In other words, in both cases, in the Minkowski space and in the Riemann space, a point-like particle moves along a geodesic.

We start with the variation of ds^2:

$$\delta ds^2 = \delta(g_{\mu\nu}dx^\mu dx^\nu) = \partial_\lambda g_{\mu\nu}\delta x^\lambda dx^\mu dx^\nu + g_{\mu\nu}(d\delta x^\mu dx^\nu + dx^\mu d\delta x^\nu).$$

By shifting d from $d\delta x^\mu$ and $d\delta x^\nu$ to other factors, i.e. in fact integrating by parts, and by changing the summation indices, we obtain

$$2ds\delta ds = \delta x^\lambda \left[(\partial_\lambda g_{\mu\nu} - \partial_\mu g_{\lambda\nu} - \partial_\nu g_{\lambda\mu})dx^\mu dx^\nu - 2g_{\lambda\nu}d^2 x^\nu\right].$$

Then, by going over to four-velocities $u^\mu = dx^\mu/ds$, we obtain in this way

$$\delta \int_A^B ds = \frac{1}{2} \int_A^B \delta x^\lambda ds \left[u^\mu u^\nu (\partial_\lambda g_{\mu\nu} - \partial_\mu g_{\lambda\nu} - \partial_\nu g_{\lambda\mu}) - 2g_{\mu\lambda}\dot{u}^\mu\right].$$

Finally, we arrive at the following equation of motion for a point-like particle in a gravitational field:

$$\frac{du^\mu}{ds} + \Gamma^\mu_{\varkappa\lambda}u^\varkappa u^\lambda = 0, \tag{2.12}$$

where

$$\Gamma^\mu_{\varkappa\lambda} = \frac{1}{2}g^{\mu\nu}(\partial_\varkappa g_{\nu\lambda} + \partial_\lambda g_{\nu\varkappa} - \partial_\nu g_{\varkappa\lambda}), \tag{2.13}$$

and the contravariant metric tensor $g^{\mu\nu}$ is related to the covariant tensor $g_{\nu\varkappa}$ as follows: $g^{\mu\nu}g_{\nu\varkappa} = \delta^\mu_\varkappa$. The quantity $\Gamma^\mu_{\varkappa\lambda}$ is called the Christoffel symbol. One can easily check that in the case of a weak gravitational field, when the metric deviates weakly from the flat one, $g_{\mu\nu} = \eta_{\mu\nu} + h_{\mu\nu}$, $|h_{\mu\nu}| \ll 1$, these equations go over into relations (2.5) and (2.6), written previously in section 2.1.

It is useful to introduce the covariant four-velocity vector $u_\mu = g_{\mu\nu}u^\nu$. Using relations (2.12) and (2.13), as well as the identity

$$\frac{dg_{\mu\nu}}{ds} = \partial_\varkappa g_{\mu\nu}\frac{dx^\varkappa}{ds} = \partial_\varkappa g_{\mu\nu}u^\varkappa,$$

one can easily demonstrate that the covariant four-velocity satisfies the equation

$$\frac{du_\mu}{ds} = \frac{1}{2}\frac{\partial g_{\nu\varkappa}}{\partial x^\mu}u^\nu u^\varkappa. \tag{2.14}$$

From it, the quite natural assertion follows: in a gravitational field independent of some coordinate x^μ, the corresponding *covariant* component of the four-velocity u_μ is conserved, and with it the *covariant* component of the four-momentum $p_\mu = mu_\mu$. For instance, in a gravitational field independent of time t, the energy $E = p_0$ is conserved, in an axially symmetric field independent of ϕ, $L_z = p_\phi$ is conserved.

A locally inertial frame at a given point corresponds to such a choice of coordinates when $g_{\mu\nu} = \eta_{\mu\nu}$, $\Gamma^{\mu}_{\varkappa\lambda} = 0$. There are many such systems, they are related to each other by Lorentz transformations.

It is sufficiently evident physically that a locally inertial frame can be chosen not only at a point, but on a geodesic as well, i.e. on the whole trajectory of a point-like particle moving in a gravitational field. Such coordinates are called normal coordinates on geodesic.

2.4 The Newton Approximation

How is equation (2.12) related to the usual equation of motion of a nonrelativistic particle in a weak gravitational field? Let the particle velocity be small, $v \ll 1$, the deviation of the metric from the flat one be small,

$$g_{\mu\nu} = \eta_{\mu\nu} + h_{\mu\nu}, \ |h_{\mu\nu}| \ll 1,$$

and in addition the fields vary slowly with time, i.e.

$$|\partial h_{\mu\nu}/\partial t| \ll |\partial h_{\mu\nu}/\partial x^m|.$$

In this approximation equation (2.12) reduces to

$$\frac{d^2 x^m}{dt^2} = -\Gamma^m_{00} = -\frac{1}{2}\,\partial_m g_{00}.$$

Now we require the validity of the Newton law:

$$\frac{d^2 x^m}{dt^2} = -\partial_m \phi\,;$$

here ϕ is the gravitational field potential. The natural boundary condition for a well-localized source of a gravitational field is:

$$g_{00} \to 1, \quad \phi \to 0 \quad \text{for} \quad |x^m| \to \infty.$$

Then

$$g_{00} = 1 + 2\phi.$$

In particular, at a large distance r from a source with a mass M we obtain

$$g_{00} = 1 - \frac{2kM}{c^2 r}$$

(we have recovered explicitly in this formula the velocity of light c).

The quantity

$$r_g = \frac{2kM}{c^2}$$

is called the gravitational radius. For the Sun (its mass $M_\odot = 2 \cdot 10^{33}$ g) the gravitational radius is $r_{g\odot} \approx 3$ km. With the radius of the Sun $R_\odot \approx 7 \cdot 10^{10}$ cm, even on its surface the deviation of the metric from the flat one is tiny: $r_{g\odot}/R_\odot \lesssim 10^{-5}$. As to the the gravitational radius of the Earth, its value is $r_{g\oplus} \approx 1$ cm.

3

Fundamentals of Riemann Geometry

3.1 Contravariant and Covariant Tensors. Tetrads

The considerations presented in the beginning of this chapter are valid for spaces more general than the space of GR. To emphasize this fact, we use here for tensor indices the Latin alphabet, but not the Greek one.

Let us consider a change of variables $x^i = f^i(x'^i)$. Under it, the differentials of coordinates transform as follows:

$$dx^i = \frac{\partial x^i}{\partial x'^k} \, dx'^k. \tag{3.1}$$

A collection of n quantities A^i, that transform under a change of coordinates as the differentials of coordinates:

$$A^i = \frac{\partial x^i}{\partial x'^k} \, A'^k, \tag{3.2}$$

is called a contravariant vector.

Let ϕ be a scalar. Its partial derivatives transform otherwise:

$$\frac{\partial \phi}{\partial x^i} = \frac{\partial x'^k}{\partial x^i} \frac{\partial \phi}{\partial x'^k}. \tag{3.3}$$

A collection of n quantities A_i transforming under a change of variables as derivatives of a scalar,

$$A_i = \frac{\partial x'^k}{\partial x^i} \, A'_k, \tag{3.4}$$

is called a covariant vector.

Tensors of higher ranks are defined in an analogous way. Thus, a contravariant tensor of second rank transforms as

$$A^{ij} = \frac{\partial x^i}{\partial x'^k} \frac{\partial x^j}{\partial x'^l} \, A'^{kl}, \tag{3.5}$$

a covariant tensor of second rank as

$$A_{ij} = \frac{\partial x'^k}{\partial x^i} \frac{\partial x'^l}{\partial x^j} A'_{kl}, \qquad (3.6)$$

a mixed tensor of second rank as

$$A^i_j = \frac{\partial x^i}{\partial x'^k} \frac{\partial x'^l}{\partial x^j} A'^k_l. \qquad (3.7)$$

Let us go back now to the interval (2.10). Since $ds^2 = g_{ij}dx^i dx^j$ is an invariant, it is clear that g_{ij} is a covariant tensor. It is called the metric tensor. The tensor g^{li} inverse to g_{ij}, i.e. related to it as

$$g^{li} g_{ij} = \delta^l_j,$$

is called a contravariant metric tensor.

Let us find now the volume element in curvilinear coordinates. We introduce vector $d\mathbf{r}$, connecting two infinitely close points x^i and $x^i + dx^i$: $d\mathbf{r} = \mathbf{e}_i dx^i$. Here $\mathbf{e}_i$ is the vector tangential to the coordinate line i going through the initial point x. It is clear that the infinitesimal vector $d\mathbf{r}$ can be described by its components dr^a in the local Lorentz (or Cartesian) frame. The expression for the vector $d\mathbf{r}$ can be rewritten as $dr^a = e^a_i dx^i$. The set of four linearly independent vectors e^a_i in a four-dimensional space, labeled by a, is called tetrad.

Obviously, the length squared of the vector $d\mathbf{r}$ is $d\mathbf{r}^2 = (\mathbf{e}_i\mathbf{e}_j)dx^i dx^j$. On the other hand, it is nothing but $ds^2 = g_{ij}dx^i dx^j$. Then it is clear that

$$g_{ij} = (\mathbf{e}_i\mathbf{e}_j) = e^a_i e_{j\,a}. \qquad (3.8)$$

It is well-known that the volume element dV, built on the vectors $\mathbf{e}_1 dx^1$, $\mathbf{e}_2 dx^2, \ldots$, is expressed via the Gram determinant:

$$dV = \sqrt{\det(\mathbf{e}_i dx^i \mathbf{e}_j dx^j)}$$

(here there is no summation over i, j), or

$$dV = \sqrt{\det(\mathbf{e}_i\mathbf{e}_j)}\, dx^1 dx^2 \ldots dx^n$$

$$= \sqrt{\det(g_{ij})}\, dx^1 dx^2 \ldots dx^n \equiv \sqrt{g}\, dx^1 dx^2 \ldots dx^n. \qquad (3.9)$$

In GR, where $g = \det(g_{ij}) < 0$, the volume element is

$$dV = \sqrt{-g}\, dx^1 dx^2 dx^3 dx^4. \qquad (3.10)$$

Problems

3.1. Is the coordinate x^i a vector?

3.2. Prove by direct calculation that $A^i B_i$ is an invariant. Prove the same for $A^{ij} B_{ij}$.

3.3. In a Euclidean space covariant tensors do not differ from contravariant ones. To what property of the rotation matrix does this coincidence correspond?

3.2 Covariant Differentiation

The differential of a vector $dA^i(x^j) = A^i(x^j + dx^j) - A^i(x^j)$ is the difference between two vectors taken at two different points. In curvilinear coordinates vectors transform in different ways at different points ($\partial x^i / \partial x'^k$ in (3.2) are functions of coordinates). Therefore, here, as distinct from the Euclidean coordinates, dA^i is not a vector. To generalize the notion of a differential dA^i in such a way as to make it a vector, one should transport at first the vector $A^i(x^j)$ parallel to itself to the point $x^j + dx^j$. Let us denote by δA^i its variation under this parallel transport. Now the difference $DA^i = dA^i - \delta A^i$ is a vector.

The variation δA^i should be linear not only in dx^j, but in A^i as well. The last point is clear from the fact that the sum of two vectors is also a vector. Thus, δA^i can be presented as

$$\delta A^i = -\,\Gamma^i_{kj}\, A^k dx^j \,, \tag{3.11}$$

where the coefficients Γ^i_{kj} are themselves functions of coordinates. In the Cartesian coordinates all $\Gamma^i_{jk} = 0$.

In line with Γ^i_{jk}, the coefficients

$$\Gamma_{l,\,jk} = g_{li}\,\Gamma^i_{jk} \tag{3.12}$$

are used.

Scalar products of vectors, as well as any scalars, do not change under the parallel transport. Then, from $\delta\,(A_i B^i) = 0$ it follows that

$$B^i \delta A_i = -A_i \delta B^i = A_i \Gamma^i_{kj} B^k dx^j \,,$$

or, since B^i are arbitrary,

$$\delta A_i = \Gamma^k_{ij} A_k dx^j \,. \tag{3.13}$$

Therefore,

$$DA^i = dA^i + \Gamma^i_{kj} A^k dx^j = \left(\partial_j A^i + \Gamma^i_{kj} A^k \right) dx^j \, ,$$

$$DA_i = dA_i - \Gamma^k_{ij} A_k dx^j = \left(\partial_j A_i - \Gamma^k_{ij} A_k \right) dx^j \, .$$

Since DA^i, DA_i and dx^j are vectors, the expressions in brackets in these equations are tensors. These tensors,

$$A^i_{;j} \equiv \frac{DA^i}{Dx^j} = \partial_j A^i + \Gamma^i_{kj} A^k, \qquad (3.14)$$

$$A_{i;j} \equiv \frac{DA_i}{Dx^j} = \partial_j A_i - \Gamma^k_{ij} A_k \, , \qquad (3.15)$$

are called covariant derivatives of the vectors A^i and A_i. Of course, in the Cartesian coordinates, where $\Gamma^i_{kj} = 0$, covariant derivatives coincide with usual ones.

Since the transformation properties of second-rank tensors are the same as those of a product of vectors, one can easily obtain the following expressions for the corresponding covariant derivatives:

$$A^{il}_{\ ;j} = \partial_j A^{il} + \Gamma^i_{kj} A^{kl} + \Gamma^l_{kj} A^{ik}; \qquad (3.16)$$

$$A^i_{l;j} = \partial_j A^i_l + \Gamma^i_{kj} A^k_l - \Gamma^k_{lj} A^i_k; \qquad (3.17)$$

$$A_{il;j} = \partial_j A_{il} - \Gamma^k_{ij} A_{kl} - \Gamma^k_{lj} A_{ik} \, . \qquad (3.18)$$

The generalization to tensors of arbitrary ranks is obvious. We note that for a scalar the covariant derivative coincides with the common one.

Since the index referring to a covariant derivative is of a tensor nature, one can raise it with the contravariant metric tensor and obtain in such a way the so-called contravariant derivative. For instance,

$$A^{i;l} = g^{lk} A^i_{\ ;k} \, , \qquad A_i^{\ ;l} = g^{lk} A_{i;k} \, .$$

How do the coefficients Γ^k_{ij} transform under a transition from one coordinate system to another? Comparing the transformation laws for the right-hand side and left-hand side of equation (3.14), we find

$$\Gamma^k_{ij} = \Gamma'^l_{mn} \frac{\partial x^k}{\partial x'^l} \frac{\partial x'^m}{\partial x^i} \frac{\partial x'^n}{\partial x^j} + \frac{\partial^2 x'^r}{\partial x^i \partial x^j} \frac{\partial x^k}{\partial x'^r} \, . \qquad (3.19)$$

It is clear now that the coefficients Γ^k_{ij} behave as tensors only under linear transformations of coordinates, just as in this case the inhomogeneous term in the right-hand side vanishes.

Let us note that this inhomogeneous term in the right-hand side of (3.19) is symmetric in i,j. Therefore, the antisymmetric in i,j combination $S^k_{ij} = \Gamma^k_{ij} - \Gamma^k_{ji}$ transforms according to

$$S_{ij}^{k} = S_{mn}^{\prime l} \frac{\partial x^k}{\partial x^{\prime l}} \frac{\partial x^{\prime m}}{\partial x^i} \frac{\partial x^{\prime n}}{\partial x^j} \, ,$$

and is thus a tensor. S_{ij}^{k} is called the torsion tensor.

In virtue of the principle of equivalence, the geometry of our space-time has a remarkable property: the torsion tensor vanishes. Indeed, in the locally inertial frame the space of GR does not differ from the flat, Minkowski one. In other words, in this frame all the coefficients $\Gamma_{\mu\nu}^{\varkappa}$, together with their antisymmetric parts S_{ij}^{k}, vanish. And since S_{ij}^{k} is a tensor, if it turns to zero in some reference frame, it vanishes identically. The spaces where the torsion tensor vanishes are called the Riemann spaces. For coordinates and tensors of a Riemann space we use the Greek indices. In a Riemann space both $\Gamma_{\nu\mu}^{\varkappa} = \Gamma_{\mu\nu}^{\varkappa}$ and $\Gamma_{\varkappa,\,\nu\mu} = \Gamma_{\varkappa,\,\mu\nu}$.

Problems

3.4. Prove relation (3.19).

3.5. How many independent components has $\Gamma_{\nu\mu}^{\varkappa}$?

3.6. Let a locally inertial frame be given at the point $x^i = 0$. Prove that under the transformation

$$x^i = x^{\prime i} + c_{jkl}^{i} x^{\prime j} x^{\prime k} x^{\prime l}$$

the frame remains locally inertial. Calculate

$$\frac{\partial \Gamma_{jk}^{\prime i}}{\partial x^{\prime l}} - \frac{\partial \Gamma_{jk}^{i}}{\partial x^l}$$

at the point $x^i = 0$.

3.3 Again Christoffel Symbols and Metric Tensor

A covariant derivative of the metric tensor vanishes. Indeed, on the one hand,

$$DA_\mu = D(g_{\mu\nu} A^\nu) = Dg_{\mu\nu} \, A^\nu + g_{\mu\nu} \, DA^\nu \, .$$

But on the other hand, as well as for any vector,

$$DA_\mu = g_{\mu\nu} \, DA^\nu \, .$$

Hence, since the vector A^ν is arbitrary,

$$Dg_{\mu\nu} = 0, \quad \text{or} \quad g_{\mu\nu;\lambda} = 0. \tag{3.20}$$

The explicit form of the last equality, with the account for (3.18), is:

$$\partial_\lambda g_{\mu\nu} - \Gamma_{\mu,\nu\lambda} - \Gamma_{\nu,\mu\lambda} = 0.$$

Interchanging indices, we obtain

$$\partial_\nu g_{\lambda\mu} - \Gamma_{\lambda,\mu\nu} - \Gamma_{\mu,\lambda\nu} = 0, \quad \partial_\mu g_{\nu\lambda} - \Gamma_{\lambda,\nu\mu} - \Gamma_{\nu,\lambda\mu} = 0.$$

Now, recalling the symmetry $\Gamma_{\mu,\nu\lambda} = \Gamma_{\mu,\lambda\nu}$, we find easily

$$\Gamma_{\lambda,\mu\nu} = \frac{1}{2}\left(\partial_\nu g_{\lambda\mu} + \partial_\mu g_{\nu\lambda} - \partial_\lambda g_{\mu\nu}\right) \tag{3.21}$$

and, correspondingly,

$$\Gamma^{\varkappa}_{\mu\nu} = \frac{1}{2} g^{\varkappa\lambda}\left(\partial_\nu g_{\lambda\mu} + \partial_\mu g_{\nu\lambda} - \partial_\lambda g_{\mu\nu}\right). \tag{3.22}$$

Thus, in a Riemann space the coefficients $\Gamma^{\varkappa}_{\mu\nu}$ coincide with the Christoffel symbols (2.13) that arise in the equations of motion of a point-like particle following from the variational principle (2.11). And this is quite natural since equation (2.12), which can be written as

$$Du^\mu = du^\mu + \Gamma^{\mu}_{\varkappa\lambda} u^{\varkappa} dx^\lambda = 0,$$

is, in accordance with the principle of equivalence, a covariant generalization for a Riemann space of the common equations of free motion

$$du^\mu = 0, \quad \text{or} \quad \frac{du^\mu}{ds} = 0.$$

We derive now a useful relation for $\Gamma^{\mu}_{\mu\nu}$. From the definition of the Christoffel symbol, it follows that

$$\Gamma^{\mu}_{\nu\mu} = \frac{1}{2} g^{\mu\lambda}\left(\partial_\nu g_{\lambda\mu} + \partial_\mu g_{\lambda\nu} - \partial_\lambda g_{\mu\nu}\right) = \frac{1}{2} g^{\mu\lambda}\,\partial_\nu g_{\lambda\mu}.$$

The metric tensor $g_{\lambda\mu}$ can be considered a matrix. Let us perform the following transformations with an arbitrary matrix M:

$$\delta\ln\det M = \ln\det(M + \delta M) - \ln\det M = \ln\det[M^{-1}(M + \delta M)]$$

$$= \ln\det(I + M^{-1}\delta M) = \ln(1 + \operatorname{Sp} M^{-1}\delta M) = \operatorname{Sp} M^{-1}\delta M.$$

Thus,

$$\operatorname{Sp} M^{-1}\partial_\nu M = \partial_\nu \ln\det M \tag{3.23}$$

and

$$\Gamma^{\mu}_{\nu\mu} = \frac{1}{\sqrt{-g}}\,\partial_\nu\sqrt{-g}\,. \tag{3.24}$$

We present two other useful relations:

$$g_{\mu\varkappa}\partial_\lambda g^{\mu\nu} = -g^{\mu\nu}\partial_\lambda g_{\mu\varkappa}; \tag{3.25}$$

$$g^{\mu\nu}\Gamma_{\mu\nu}^{\varkappa} = -\frac{1}{\sqrt{-g}}\,\partial_\mu(\sqrt{-g}\,g^{\varkappa\mu})\,. \tag{3.26}$$

The covariant generalization of the divergence of a vector is:

$$A^\mu_{;\mu} = \frac{1}{\sqrt{-g}}\,\partial_\mu(\sqrt{-g}\,A^\mu)\,. \tag{3.27}$$

It follows in particular from the last equation that in a Riemann space the application of the Dalembert operator to a scalar ϕ is as follows:

$$\phi^{;\mu}_{\ ;\mu} = \frac{1}{\sqrt{-g}}\,\partial_\mu(\sqrt{-g}\,g^{\mu\nu}\partial_\nu\phi)\,. \tag{3.28}$$

The Gauss theorem is now

$$\int d^4x\,\sqrt{-g}\,A^\mu_{;\mu} = \oint dS_\mu\,\sqrt{-g}\,A^\mu\,. \tag{3.29}$$

One more useful relation is:

$$A_{\mu;\nu} - A_{\nu;\mu} = \partial_\nu A_\mu - \partial_\mu A_\nu\,. \tag{3.30}$$

The covariant divergence of an antisymmetric tensor $A^{\mu\nu} = -A^{\nu\mu}$ is

$$A^{\mu\nu}_{\ ;\nu} = \frac{1}{\sqrt{-g}}\,\partial_\nu(\sqrt{-g}\,A^{\mu\nu})\,. \tag{3.31}$$

Besides, we have for this tensor

$$A_{\mu\nu;\lambda} + A_{\nu\lambda;\mu} + A_{\lambda\mu;\nu} = \partial_\lambda A_{\mu\nu} + \partial_\mu A_{\nu\lambda} + \partial_\nu A_{\lambda\mu}\,. \tag{3.32}$$

Problems

3.7. Prove formulae (3.25) – (3.32).

3.8. Is $A = \det(A_{\mu\nu})$ a scalar? Here $A_{\mu\nu}$ is a second-rank tensor.

3.9. Calculate the Christoffel symbols for cylindrical and spherical coordinates.

3.10. Present the explicit form of formulae (3.27) and (3.28) in cylindrical and spherical coordinates.

3.11. Write the Maxwell equations in a Riemann space.

3.4 Simple Illustration of Some Properties of Riemann Space

A transparent intuitive idea of some properties of a Riemann space can be given with the simplest example of a sphere. Let us consider on it a spherical triangle, whose sides are arcs of great circles. An arc of a great circle connecting two points on a sphere is known to be the shortest path between them, i.e. this is a geodesic. Here we choose as these arcs those of the meridians, differing by 90° of longitude, and of the equator (see Fig. 3.1). The sum of the angles of this triangle in no way coincides with π, the sum of the angles of a triangle

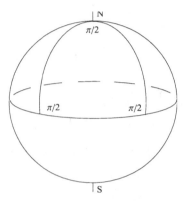

Fig. 3.1. Spherical triangle

on a plane, but equals to

$$\alpha + \beta + \gamma = \frac{3}{2}\pi. \tag{3.33}$$

We note that the excess of this sum of the angles over π can be expressed via the area S of the triangle and the radius R of the sphere:

$$\alpha + \beta + \gamma - \pi = \frac{S}{R^2}. \tag{3.34}$$

This relation can be demonstrated to hold for any spherical triangle. We note as well that the common case of a triangle on a plane follows also from this formula: a plane can be considered as a sphere with $R \to \infty$.

Let us rewrite formula (3.34) otherwise:

$$K \equiv \frac{1}{R^2} = \frac{\alpha + \beta + \gamma - \pi}{S}. \tag{3.35}$$

It is clear from this relation that one can determine the radius of a sphere while being confined to it, without going from the sphere to the three-dimensional space into which the sphere is embedded. To this end, it is sufficient to measure the area of a spherical triangle and the sum of its angles. In other words, R and K are in fact internal characteristics of a sphere. The quantity K, which is called the Gauss curvature, is generalized in a natural way to an arbitrary smooth surface:

$$K(x) = \lim_{S \to 0} \frac{\alpha + \beta + \gamma - \pi}{S}. \qquad (3.36)$$

Here the angles and area refer to a small triangle on the surface, bounded by geodesics on it, and the curvature is a local characteristic that changes generally speaking from point to point. In a general case, as well as for a sphere, K is an internal characteristic of a surface, independent of its embedding into the three-dimensional space. The Gauss curvature of a surface does not change under bending of a surface without tearing or stretching it. So, for instance, a cylinder can be unbent into a plane, and thus for it $K = 0$, as well as for a plane.

It is instructive to look at relations (3.35) and (3.36) somewhat otherwise. Let us go back to Fig. 3.1. We take at the pole a vector directed along one of the meridians, and transfer it along this meridian, without changing the angle between the vector and the meridian (which is zero in the present case), to the equator. Then we transfer the vector along the equator, again without changing the angle between them (which is $\pi/2$ now), to the second meridian. And at last, we come back in the same way along the second meridian to the pole. It can be easily seen that, as distinct from the same transport along a closed path on a plane, the vector will be finally rotated with respect to its initial direction by $\pi/2$, or by

$$\alpha + \beta + \gamma - \pi = KS. \qquad (3.37)$$

This result, rotation of a vector under its parallel transport along a closed path by an angle proportional to the area inside the contour, is generalized in a natural way not only to an arbitrary two-dimensional surface, but to multidimensional non-Euclidean spaces as well. However, in the general case of an n-dimensional space the curvature does not reduce to a single scalar function $K(x)$. It is now a more complicated geometrical object — the curvature tensor, or the Riemann tensor. That is what we will now examine.

3.5 Tensor of Curvature

If $x^\mu(s)$ is a parametric equation of a curve (here s is the distance along it), then $u^\mu = dx^\mu/ds$ is the unit vector tangential to the curve. If this curve is a geodesic, then along it $Du^\mu = 0$. In other words, if u^μ is parallel transported from the point x^μ on the geodesic to the point $x^\mu + dx^\mu$ on it, then it will coincide with the unit vector $u^\mu + du^\mu$, tangential to the geodesic at the point $x^\mu + dx^\mu$. Thus, under the motion along a geodesic the tangential unit vector is transported along itself.

By definition, under a parallel transport of two vectors the "angle" between them remains constant. Therefore, under a parallel transport of any vector along a geodesic, the angle between it and the vector tangential to the geodesic does not change, i.e. the projections of the transported vector onto geodesic lines at all points of the path remain constant.

We have seen already that a vector on the surface of a sphere does not coincide with itself at the initial point after a parallel transport along a closed contour. Now we will consider a more general problem: we will find the change ΔA_μ of a vector A_μ under a parallel transport along an infinitesimal closed contour in a Riemann space. In the general case this change is written as the integral $\oint \delta A_\mu$ along the contour. With the account for (3.13), we obtain

$$\Delta A_\mu = \oint \delta A_\mu = \oint \Gamma^\nu_{\mu\lambda} A_\nu dx^\lambda. \tag{3.38}$$

We transform this integral by means of the Stokes theorem. To this end we need the values of the vector A_μ inside the infinitesimal contour of integration. Strictly speaking, these values are not functions of a point, but depend themselves on the path by which this point is reached. However, for an infinitesimal contour this ambiguity is an infinitesimal quantity of second order. Thus, one can neglect the ambiguity and define the vector A_μ inside the contour via its values on the contour itself, by means of derivatives:

$$\partial_\sigma A_\nu = \Gamma^\rho_{\nu\sigma} A_\rho. \tag{3.39}$$

Now, recalling again that the area $\Delta f^{\rho\tau}$ inside the contour is infinitesimal, we obtain with the Stokes theorem

$$\Delta A_\nu = \frac{1}{2} \left[\partial_\rho (\Gamma^\mu_{\nu\tau} A_\mu) - \partial_\tau (\Gamma^\mu_{\nu\rho} A_\mu) \right] \Delta f^{\rho\tau}$$

$$= \frac{1}{2} \left[\partial_\rho \Gamma^\mu_{\nu\tau} A_\mu - \partial_\tau \Gamma^\mu_{\nu\rho} A_\mu + \Gamma^\mu_{\nu\tau} \partial_\rho A_\mu - \Gamma^\mu_{\nu\rho} \partial_\tau A_\mu \right] \Delta f^{\rho\tau}.$$

With the account for (3.39), we obtain finally

$$\Delta A_\nu = \frac{1}{2} R^\mu{}_{\nu\rho\tau} A_\mu \Delta f^{\rho\tau}, \tag{3.40}$$

where

$$R^\mu{}_{\nu\rho\tau} = \partial_\rho \Gamma^\mu_{\nu\tau} - \partial_\tau \Gamma^\mu_{\nu\rho} + \Gamma^\mu_{\sigma\rho} \Gamma^\sigma_{\nu\tau} - \Gamma^\mu_{\sigma\tau} \Gamma^\sigma_{\nu\rho} \tag{3.41}$$

is the curvature tensor, or the Riemann tensor.

An analogous formula is valid also for a covariant vector A^ν. Since scalars do not change under a covariant transport, we have

$$\Delta(A^\nu B_\nu) = \Delta A^\nu B_\nu + A^\nu \Delta B_\nu = \Delta A^\nu B_\nu + A^\nu \frac{1}{2} R^\mu{}_{\nu\rho\tau} B_\mu \Delta f^{\rho\tau}$$

$$= B_\mu (\Delta A^\mu + \frac{1}{2} R^\mu{}_{\nu\rho\tau} A^\nu \Delta f^{\rho\tau}) = 0.$$

Since the vector B_μ is arbitrary, it means that

$$\Delta A^\mu = -\frac{1}{2} R^\mu{}_{\nu\rho\tau} A^\nu \Delta f^{\rho\tau}. \qquad (3.42)$$

The operations of the covariant differentiation do not commute. In particular,

$$A_{\rho;\,\mu;\,\nu} - A_{\rho;\,\nu;\,\mu} = R^\tau{}_{\rho\mu\nu} A_\tau, \qquad (3.43)$$

$$A^\rho{}_{;\,\mu;\,\nu} - A^\rho{}_{;\,\nu;\,\mu} = -R^\rho{}_{\tau\mu\nu} A^\tau, \qquad (3.44)$$

$$A_{\rho\sigma;\,\mu;\,\nu} - A_{\rho\sigma;\,\nu;\,\mu} = R^\tau{}_{\rho\mu\nu} A_{\tau\sigma} + R^\tau{}_{\sigma\mu\nu} A_{\rho\tau}. \qquad (3.45)$$

These tensor relations can be easily proven in a locally inertial frame.

In a flat space the Riemann tensor vanishes. Indeed, in such a space one can choose the coordinates in such a way that $\Gamma^\mu_{\nu\rho} = 0$ everywhere, and hence $R^\tau{}_{\rho\mu\nu} = 0$. And a tensor vanishing in one coordinate frame, vanishes in any other one.

The inverse statement is also true: if the Riemann tensor vanishes, the space is flat. Indeed, locally, at a given point one can choose a Euclidean frame in any space. And with $R^\tau{}_{\rho\mu\nu} = 0$ the parallel transport of the Euclidean coordinate frame from a given point to any other one is path-independent. Thus, the Euclidean frame can be built in a unique way in the whole space. And this means in fact that the space is flat.

Problems

3.12. Prove formulae (3.43) – (3.45).

3.13. What is the form of the Dalembert equation (2.4) in a gravitational field in the covariant Lorentz gauge where $A^\mu{}_{;\,\mu} = 0$?

3.14. In the flat space-time the electromagnetic field strength $F_{\mu\nu}$ satisfies the equation $\Box F_{\mu\nu} = 0$. What is the form of the corresponding equation in a gravitational field?

3.6 Properties of the Riemann Tensor

The antisymmetry of the Riemann tensor in the last two indices,

$$R^\tau{}_{\rho\mu\nu} = -R^\tau{}_{\rho\nu\mu},$$

is obvious from its definition (see (3.40) and (3.41)). To investigate its other symmetry properties, it is convenient to go over from the mixed components to covariant ones:

$$R_{\tau\rho\mu\nu} = g_{\tau\sigma} R^{\sigma}{}_{\rho\mu\nu} \, .$$

Going over again into the locally inertial frame, one can prove the following symmetry properties of the tensor $R_{\tau\rho\mu\nu}$:

$$R_{\tau\rho\mu\nu} = -R_{\rho\tau\mu\nu} \, , \tag{3.46}$$

$$R_{\tau\rho\mu\nu} = R_{\mu\nu\tau\rho} \, . \tag{3.47}$$

The antisymmetry in the first two indices of (3.46) is sufficiently obvious: it guarantees the conservation of the length of a vector under its transport along a closed contour. Less obvious is the symmetry under the permutation of the pairs of indices in (3.47), since the meaning of these pairs is different. The first one refers to the vector we transport, and the second refers to the site around which this vector is transported.

Then, the cyclic sum of the Riemann tensor components over three indices, with the fourth one fixed, vanishes:

$$R_{\tau\rho\mu\nu} + R_{\tau\mu\nu\rho} + R_{\tau\nu\rho\mu} = 0 \, . \tag{3.48}$$

And at last, there is the Bianchi identity:

$$R^{\sigma}{}_{\rho\mu\nu;\,\tau} + R^{\sigma}{}_{\rho\tau\mu;\,\nu} + R^{\sigma}{}_{\rho\nu\tau;\,\mu} = 0. \tag{3.49}$$

By contracting the Riemann tensor in two indices one obtains a second-rank tensor, or the Ricci tensor. We define it as follows:

$$R_{\mu\nu} = R^{\rho}{}_{\mu\rho\nu} = \partial_{\rho}\Gamma^{\rho}_{\mu\nu} - \partial_{\nu}\Gamma^{\rho}_{\mu\rho} + \Gamma^{\rho}_{\sigma\rho}\Gamma^{\sigma}_{\mu\nu} - \Gamma^{\rho}_{\sigma\nu}\Gamma^{\sigma}_{\mu\rho} \, . \tag{3.50}$$

Any other contraction of the curvature tensor either turns to zero or coincides with this one up to the sign. The Ricci tensor is symmetric:

$$R_{\mu\nu} = R_{\nu\mu} \, . \tag{3.51}$$

The contraction of the Ricci tensor gives an invariant — the scalar curvature of the space

$$R = g^{\mu\nu} R_{\mu\nu} \, . \tag{3.52}$$

We point out also the differential identity

$$R^{\nu}_{\mu;\,\nu} = \frac{1}{2} \partial_{\mu} R \, , \tag{3.53}$$

that arises under contracting the Bianchi identity (3.49).

Let us find the number of independent components of the Riemann tensor for a space of an arbitrary dimension n. The tensor $R_{\tau\rho\mu\nu}$ is antisymmetric under the permutations $\tau \longleftrightarrow \rho$, $\mu \longleftrightarrow \nu$. Therefore, the total number of independent combinations in an n-dimensional space for both pairs $\tau\rho$ and $\mu\nu$ is $n(n-1)/2$. On the other hand, the tensor $R_{\tau\rho\mu\nu}$ is symmetric under the permutation of these pairs, $\tau\rho \longleftrightarrow \mu\nu$. Hence the total number of independent combinations of the indices is

$$\frac{1}{2}\frac{n(n-1)}{2}\left[\frac{n(n-1)}{2}+1\right].$$

However, one should also take into account the cyclic conditions (3.48):

$$B_{\tau\rho\mu\nu} = R_{\tau\rho\mu\nu} + R_{\tau\mu\nu\rho} + R_{\tau\nu\rho\mu} = 0\,.$$

To find the number of them, note that the tensor $B_{\tau\rho\mu\nu}$ is totally antisymmetric. For instance,

$$B_{\rho\tau\mu\nu} = R_{\rho\tau\mu\nu} + R_{\rho\mu\nu\tau} + R_{\rho\nu\tau\mu} = -R_{\tau\rho\mu\nu} - R_{\tau\mu\nu\rho} - R_{\tau\nu\rho\mu} = -B_{\tau\rho\mu\nu}\,.$$

It can be easily seen therefore that the total number of independent cyclic conditions (3.48) is $n(n-1)(n-2)(n-3)/4!$. Finally, the total number of independent components of the Riemann tensor is

$$\frac{1}{2}\frac{n(n-1)}{2}\left[\frac{n(n-1)}{2}+1\right] - \frac{n(n-1)(n-2)(n-3)}{4!}$$

$$= \frac{n^2(n^2-1)}{12}\,. \tag{3.54}$$

In particular, the numbers of independent components of the Riemann tensor are: 20 for $n = 4$, 6 for $n = 3$, and 1 for $n = 2$.

However, the number of independent components of the curvature tensor at any given point can be made even smaller. Indeed, the locally inertial (or locally Euclidean) system at a given point is defined up to rotations. By a corresponding choice of rotation parameters one can turn to zero $n(n-1)/2$ components more of the curvature tensor. As a result, the curvature of a four-dimensional space is characterized at any point by 14 quantities, and that of a three-dimensional one — by 3 quantities. This consideration does not apply to two dimensions, where one can choose as the only characteristic the scalar curvature: a scalar cannot be turned to zero by any rotations.

In a four-dimensional space, under the condition $R_{\mu\nu} = 0$ (it will be demonstrated in the next chapter that this is the property of the Riemann tensor in an empty space), the curvature tensor has 10 independent components. For any given point of this space the coordinate frame can be chosen in such a way that all the components of $R_{\tau\rho\mu\nu}$ are expressed via no more than 4 independent quantities.

Problems

3.15. Prove formulae (3.46) – (3.49).

3.16. Express the Riemann tensor in a two-dimensional space via the scalar curvature.

3.17. Express the Riemann tensor in a three-dimensional space via the scalar curvature and the Ricci tensor.

3.18. How is the scalar curvature of a sphere related to the radius of this sphere?

3.19. Calculate the Riemann tensor, the Ricci tensor, and the scalar curvature of the surface of a torus.

3.20. Calculate the Riemann tensor of the surface of a cone. Investigate the integral $\int \sqrt{g}\, d^2x\, R$ near the top of the cone as follows: approximate the top of the cone by a spherical cap and then let the radius r of the cap tend to zero.

3.21. Choose a locally inertial frame at some point, with this point taken as the origin. Prove that the metric tensor in the vicinity of this point can be expressed through the Riemann tensor as follows:

$$g_{\mu\nu} = \eta_{\mu\nu} - \frac{1}{3} R_{\mu\alpha\nu\beta} x^\alpha x^\beta.$$

3.7 Relative Acceleration of Two Particles Moving Along Close Geodesics

Let a particle a move in a gravitational field. In the normal coordinates on its geodesic, the motion of this particle is free:

$$\frac{d^2 x_a^\mu}{ds^2} = 0.$$

The equation of motion of a particle b moving along a neighboring geodesic,

$$\frac{d^2 x_b^\mu}{ds^2} + \Gamma^\mu_{\rho\tau}(x_b) \frac{dx_b^\rho}{ds} \frac{dx_b^\tau}{ds} = 0,$$

reduces, to first order in the difference of the coordinates $\eta^\mu(s) = x_b^\mu(s) - x_a^\mu(s)$ (this difference is called geodesic deviation), to

$$\frac{d^2 x_b^\mu}{ds^2} + \partial_\nu \Gamma^\mu_{\rho\tau}(x_a) \eta^\nu \frac{dx_a^\rho}{ds} \frac{dx_a^\tau}{ds} = 0.$$

Then in the normal coordinates on the geodesic of the particle a, the equation for the geodesic deviation η^μ is

$$\frac{d^2 \eta^\mu}{ds^2} + \partial_\nu \Gamma^\mu_{\rho\tau} \eta^\nu u^\rho u^\tau = 0. \tag{3.55}$$

This equation can be rewritten in a covariant form valid in an arbitrary reference frame. We note to this end that the usual derivative of any order along a geodesic coincides with the covariant one, so that one may write $D^2\eta^\mu/Ds^2$ instead of $d^2\eta^\mu/ds^2$. Then, in the normal coordinates the Christoffel symbol on a geodesic vanishes, so that

$$\partial_\tau \Gamma^\mu_{\rho\nu} u^\tau = \frac{d\Gamma^\mu_{\rho\nu}}{ds} = 0 \,.$$

Hence in the second term in (3.55) one may substitute

$$\partial_\nu \Gamma^\mu_{\rho\tau} u^\tau \rightarrow (\partial_\nu \Gamma^\mu_{\rho\tau} - \partial_\tau \Gamma^\mu_{\rho\nu}) u^\tau \,.$$

The last expression is written in the normal coordinates, and its covariant form is $R^\mu_{\ \rho\nu\tau} u^\tau$. In result, we arrive at the following generally covariant equation of the geodesic deviation:

$$\frac{D^2\eta^\mu}{Ds^2} + R^\mu_{\ \rho\nu\tau} u^\rho u^\tau \eta^\nu = 0 \,. \tag{3.56}$$

This equation describes in fact the tidal forces acting on a system of two particles in an inhomogeneous gravitational field.

4

Einstein Equations

4.1 General Form of Equations

It is natural to assume that the generally covariant equations of the gravitational field should be second-order differential equations, and that the energy-momentum tensor $T^{\mu\nu}$ should serve as a source in them. An additional assumption is that these equations should be linear in the Riemann tensor. Then their general structure is

$$aR^{\mu\nu} + bg^{\mu\nu}R + cg^{\mu\nu} = T^{\mu\nu}.$$

The condition $T^{\mu\nu}{}_{;\nu} = 0$ and identity (3.53) dictate that $b = -a/2$. In this way we arrive at the Einstein equations

$$R^{\mu\nu} - \frac{1}{2}\,g^{\mu\nu}R = 8\pi k T^{\mu\nu} + \Lambda g^{\mu\nu}. \tag{4.1}$$

The coefficient $8\pi k$ at $T^{\mu\nu}$ (k is the Newton constant) guarantees, as will be demonstrated below, the agreement with the common Newton law in the corresponding approximation. The so-called cosmological constant Λ is at any rate extremely small, according to experimental data; therefore, the last term in the left-hand side of equation (4.1) is usually omitted.

We note that if nevertheless $\Lambda \neq 0$, the cosmological term in (4.1) can be presented as an effective additional contribution

$$\tau^{\mu\nu} = \frac{\Lambda}{8\pi k}\,g^{\mu\nu}$$

to the energy-momentum tensor of the matter $T^{\mu\nu}$. This contribution is quite peculiar. As distinct from the energy-momentum tensor of particles with a rest mass, for $\tau^{\mu\nu}$ there is no reference frame where only the component τ^{00} differs from zero. As distinct from the energy-momentum tensor of massless particles, the trace of $\tau^{\mu\nu}$ does not vanish: $\tau^\mu_\mu = \Lambda/2\pi k$.

On the other hand, in the locally geodesic frame

$$\tau^{\mu\nu} = \frac{\Lambda}{8\pi k}\,\eta^{\mu\nu} = \frac{\Lambda}{8\pi k}\,\mathrm{diag}(1,-1,-1,-1)\,.$$

With this diagonal tensor $\tau^{\mu\nu}$, the corresponding effective energy density ρ_Λ and pressure p_Λ are as follows[1]:

$$\tau^{\mu\nu} = \mathrm{diag}(\rho_\Lambda, p_\Lambda, p_\Lambda, p_\Lambda)\,. \tag{4.2}$$

Clearly, such a peculiar "matter" has also quite a peculiar equation of state:

$$p_\Lambda = -\rho_\Lambda = -\tau_{00} = -\frac{\Lambda}{8\pi k}\,, \tag{4.3}$$

i.e. its pressure is negative! Modern data of the observational astronomy give serious reasons to believe that the cosmological term does not vanish. It is quite possible that, though being tiny on the usual scale, the cosmological term is very essential for the evolution of the Universe.

In the absence of matter $T^{\mu\nu} = 0$ and the Einstein equations (4.1) reduce to

$$R^{\mu\nu} = 0\,. \tag{4.4}$$

The spaces with metric satisfying condition (4.4) are called the Einstein spaces. Equation (4.1) (in the absence of the cosmological constant) can be rewritten as:

$$R^{\mu\nu} = 8\pi k\left(T^{\mu\nu} - \frac{1}{2}\,g^{\mu\nu}T^\lambda_\lambda\right)\,. \tag{4.5}$$

The Einstein equations are in essence the content of general relativity.

Problem

4.1. Prove relation

$$R^{\alpha\beta\mu\nu}{}_{;\nu} = 8\pi k\left[T^{\mu\alpha;\beta} - T^{\mu\beta;\alpha} - \frac{1}{2}\left(g^{\mu\alpha}T^{\lambda;\beta}_\lambda - g^{\mu\beta}T^{\lambda;\alpha}_\lambda\right)\right]$$

(*A. Lichnerowicz*, 1960).

4.2 Linear Approximation

In the linear approximation, $g_{\mu\nu} = \eta_{\mu\nu} + h_{\mu\nu}$, $|h_{\mu\nu}| \ll 1$, the Ricci tensor is

$$R_{\mu\nu} = \frac{1}{2}\,[\partial_\rho\partial_\nu h_{\mu\rho} + \partial_\mu\partial_\rho h_{\nu\rho} - \Box h_{\mu\nu} - \partial_\mu\partial_\nu h_{\rho\rho}]\,.$$

---

[1]See, for instance, L.D. Landau and E.M. Lifshitz, *The Classical Theory of Fields*, §35, formula (35.1).

We use for the metric the gauge

$$\partial_\mu h_{\mu\nu} - \frac{1}{2}\partial_\nu h_{\mu\mu} = 0\,, \qquad (4.6)$$

analogous to the Lorentz condition $\partial_\mu A_\mu = 0$ in electrodynamics. In this gauge the Einstein equation reduces in the linear approximation to the usual wave equation (of course, for a massless field)

$$-\Box h_{\mu\nu} = 16\pi k\left(T_{\mu\nu} - \frac{1}{2}\eta_{\mu\nu}T_{\lambda\lambda}\right). \qquad (4.7)$$

As well as in electrodynamics, in the linear approximation there is no real difference between upper and lower indices.

Let us consider the case when the source of the field is a body at rest with density ρ, i.e. when the only nonvanishing component of the energy-momentum tensor is $T_{00} = \rho$. Then

$$\Delta h_{00} = 8\pi k\rho$$

and

$$h_{00}(\mathbf{r}) = -2k\int \frac{\rho(\mathbf{r}')d\mathbf{r}'}{|\mathbf{r} - \mathbf{r}'|}\,.$$

Thus, at large distances from the gravitating mass M we find, as expected,

$$h_{00} = -\frac{2k}{r}\int \rho(\mathbf{r}')d\mathbf{r}' = -\frac{2kM}{r}\,. \qquad (4.8)$$

In this gauge, other components of the metric far from the gravitating mass are

$$h_{0n} = 0, \qquad h_{mn} = -\frac{2kM}{r}\delta_{mn}\,. \qquad (4.9)$$

Of course, equation (4.7) has nontrivial wave solutions even in the absence of sources. The existence of gravitational waves is an important prediction of general relativity.

4.3 Again Electrodynamics and Gravity

In section 2.1. we pointed out some similarity between electrodynamics and gravity. Now we wish to turn attention to an essential difference between them. It is well known that the Maxwell equations result in only one scalar condition, that of the electromagnetic current conservation. In no way does the vector equation of motion of the charge, which has four components, follow from them. Indeed, when applying to $\partial_\mu F^{\mu\nu} = 4\pi j^\nu$ the operator ∂_ν, we obtain $\partial_\nu j^\nu = 0$. This single scalar continuity equation tells us not so much about the motion of the charged particle: only that its world-line does not break anywhere.

Now, by applying the covariant derivative D/Dx^ν to the Einstein equation (4.1), we arrive at the vector equation

$$T^{\mu\nu}{}_{;\nu} = 0\,. \tag{4.10}$$

As distinct from the current conservation law, the four equations (4.10) ($\mu = 0, 1, 2, 3$ therein) determine completely the motion of particles. Let us demonstrate it with the example of dust, i.e. a cloud of point-like noninteracting particles of small mass, moving in an external gravitational field. The energy-momentum tensor of dust is $T^{\mu\nu} = \rho u^\mu u^\nu$, where ρ is the invariant energy density initially defined in the comoving frame. Equation (4.10) can be rewritten here as follows:

$$T^{\mu\nu}{}_{;\nu} = (\rho u^\mu u^\nu)_{;\nu} = (\rho u^\nu)_{;\nu} u^\mu + \rho u^\mu{}_{;\nu} u^\nu = (\rho u^\nu)_{;\nu} u^\mu + \rho \frac{Du^\mu}{Ds} = 0\,. \tag{4.11}$$

Multiplying the obtained identity by u_μ and taking into account that $u^\mu u_\mu = 1$ and therefore $u_\mu Du^\mu/Ds = 0$, we obtain first of all the continuity equation for the current density of the dust particles

$$(\rho u^\nu)_{;\nu} = 0\,,$$

and then the required equation of motion

$$\frac{Du^\mu}{Ds} = 0\,.$$

The example of dust was chosen for simplicity sake only. For a single particle as well one can prove that its equations of motion are contained in the Einstein equations.

This remarkable property of the equations of gravity was formulated by Einstein as follows: "Matter dictates to space how to bend; space dictates to matter how to move."

As to electrodynamics, its equations are linear, the superposition principle is valid therein, the sum of the fields of particles at rest is the solution as well as the field of each of them. Therefore, if the equations of motion of charged particles in the electromagnetic field were not given, the charges initially at rest could stay at rest further. But since the equations of GR are nonlinear, there is no superposition principle here, so that bodies initially at rest should start moving. In fact, this argument is closely related to the above derivation of the equations of motion, based on the existence of four conservation laws for the tensor equations of gravitational field. The point is that the nonlinearity of the field equations is an inevitable consequence of their tensor structure.

4.4 Are Alternative Theories of Gravity Viable?

First of all, the long-range nature of gravity is firmly established, so that it should be described by a massless field (or at least the rest mass of this field should be extremely small).

The simplest alternative to the Einstein gravity, one could think about, is a scalar theory. The relativistic invariance demands that the scalar field should interact with a scalar characteristic of matter. Such a reasonable characteristic is the trace T_μ^μ of its energy-momentum tensor. However, for massless particles, light included, $T_\mu^\mu = 0$. Thus, in a scalar theory light will not interact with a gravitational field. However, the light deflection by the gravitational field of the Sun, the retardation of light in this field, as well as the frequency shift by the gravitational field of the Earth are firmly established experimental facts.

The situation with a vector theory is no better. The interactions of particles and antiparticles with the vector field (as well as in the common electrodynamics) have opposite signs. But certainly it is not so. Besides, here as well the neutral photon will not interact with a gravitational field.

To summarize, general relativity, where the gravitational field is described by a symmetric second-rank tensor, is the simplest theory of gravity consistent with experiment.

With the best accuracy, of about 0.2%, the predictions of GR have been checked experimentally for the retardation of light in the field of the Sun (see section 6.5). Strictly speaking, one cannot exclude that on this level there is an admixture of a scalar field to the tensor one.

5

Weak Field. Observable Effects

5.1 Shift of Light Frequency
in Constant Gravitational Field

We start with an estimate for the possible magnitude of the effect. If the gravitational field of the Earth is meant, then it is quite natural to assume that the frequency shift of light $\Delta\omega/\omega$, as measured by a detector situated at the height h above the source, should be proportional to this height as well as to the free-fall acceleration g. Then simple dimensional arguments give

$$\frac{\Delta\omega}{\omega} \sim \frac{gh}{c^2},$$

where c is the velocity of light.

And now the quantitative consideration. In a constant field (i.e. independent of the world time t) the energy E is conserved. It is well-known to be related to the action S as follows: $E = -\partial S/\partial t$. Exactly in the same way, in a constant field the wave frequency ω is conserved, and it is related to the eikonal Ψ as follows:

$$\omega = -\frac{\partial\Psi}{\partial t}.$$

However, both the clock that is at rest together with the source of light, and the clock that is at rest together with the detector of light show the proper time, each one its own. The frequency in the proper time τ

$$\omega_\tau = -\frac{\partial\Psi}{\partial\tau} = -\frac{\partial\Psi}{\partial t}\frac{\partial t}{\partial\tau} = \frac{\omega}{\sqrt{g_{00}}}$$

in the weak gravitational field of the Earth reduces to

$$\omega_\tau = \omega\left(1 + \frac{kM}{r}\right).$$

If the detector is situated at the height h over the source, then the frequency fixed by the detector will be red-shifted as compared to the frequency of the source. This shift is ($A.$ $Einstein$, 1907)

$$\omega_\tau(r + h) - \omega_\tau(r) = -\omega\,\frac{kMh}{r^2} = -\omega\,\frac{gh}{c^2}\,.$$

In the final expression we have recovered explicitly the velocity of light c. The agreement with the initial simple-minded estimate is obvious.

The relative magnitude of the correction is extremely tiny. Even for $h \sim$ 100 m it is

$$\frac{\Delta\omega}{\omega} \sim 10^{-14}\,.$$

For the first time, the effect was measured in the Mössbauer transition in ^{57}Fe. The theoretical prediction is confirmed within the experimental error that is about 1%.

5.2 Light Deflection by the Sun

An obvious dimensional estimate for the deflection angle θ is

$$\theta \sim \frac{r_g}{\rho}\,,$$

where ρ is the impact parameter of the wave packet. The result $\theta = r_g/\rho$ follows also from the naive calculation based on the picture of a fast particle scattered by a small angle by the usual Newton potential.

The weak-field approximation is quite sufficient for the quantitative calculation of the discussed effect. In this approximation the generally covariant eikonal equation

$$g^{\mu\nu}\partial_\mu\Psi\,\partial_\nu\Psi = 0$$

reduces in the centrally symmetric field to

$$\left(1 + \frac{r_g}{r}\right)(\partial_t\Psi)^2 - \left(1 - \frac{r_g}{r}\right)\left[(\partial_r\Psi)^2 + \frac{1}{r^2}\,(\partial_\phi\Psi)^2\right] = 0\,. \tag{5.1}$$

We use here the solution (4.8) and (4.9) for the metric far away from the gravitating mass; in this approximation the nonvanishing contravariant components of the metric are

$$g^{00} = 1 + \frac{r_g}{r}\,, \quad g^{mn} = -\delta_{mn}\left(1 - \frac{r_g}{r}\right)\,. \tag{5.2}$$

Then we go over to the spherical coordinates and assume that the motion takes place in the plane $\theta = \pi/2$.

For small r_g/r, equation (5.1) is conveniently rewritten as follows:

$$\left(1 + 2\,\frac{r_g}{r}\right)(\partial_t\Psi)^2 - \left[(\partial_r\Psi)^2 + \frac{1}{r^2}\,(\partial_\phi\Psi)^2\right] = 0\,. \tag{5.3}$$

We look for the solution in the form

$$\Psi = -\omega t + \omega\rho\phi + \psi(r)\,,$$

where ω is the frequency of light. The correspondence of the impact parameter ρ to the common integral L of the orbital angular momentum is obvious: $\rho \to L/\hbar\omega$ (we put here the velocity of light $c = 1$).

The radial part of the eikonal is

$$\psi(r) = \omega\int dr\,\sqrt{1 - \frac{\rho^2}{r^2} + \frac{2r_g}{r}} = \psi_0(r) + \Delta\psi(r).$$

Here $\psi_0(r)$ describes the unperturbed rectilinear motion of the packet, and the small gravitational correction to it is

$$\Delta\psi(r) = \omega r_g\int\frac{dr}{\sqrt{r^2 - \rho^2}} = \omega r_g\ln\left(r + \sqrt{r^2 - \rho^2}\right) + \text{const.}$$

As usual, the trajectory of the packet is found by differentiating the total eikonal over the integral of motion:

$$\frac{\partial\Psi}{\partial\rho} = \text{const}, \qquad \phi = -\frac{1}{\omega}\,\frac{\partial\psi}{\partial\rho}\,.$$

Thus obtained deviation of the beam of light from the straight line, when its distance r to the Sun changes from $-R$ to ρ, and then from ρ to R $(R \to \infty)$, is

$$\theta = -\frac{1}{\omega}\,\frac{\partial\Delta\psi}{\partial\rho} = -2r_g\,\frac{\partial}{\partial\rho}\ln\frac{2R}{\rho} = \frac{2r_g}{\rho}\,. \tag{5.4}$$

For the minimum ρ close to the Sun radius, the deflection angle θ is $1.75''$. This prediction of GR (*A. Einstein*, 1915) is confirmed now by observations with an accuracy of about 1%.

Let us recall that the naive calculation of the effect, based on the picture of a fast particle deflected by a small angle in the usual Newton potential, gives a result (see the beginning of the section) that is two times smaller than the correct one. The discrepancy is no occasion: in the considered ultrarelativistic problem not only the Newton potential is at work, i.e. the deviation of g_{00} from unit. Exactly the same contribution to the deflection is given by the space metric g_{mn} (see (5.1) – (5.3)).

5.3 Gravitational Lenses

Since a star deflects rays of light, it can be considered as a peculiar gravitational lens. Such a lens shifts the image of a source (i.e. of a star) with respect to its true position. In the simplest case, when the source, lens, and observer are on the same axis, the image of the source looks as a circle (*O.D. Chwolson*, 1924; *A. Einstein*, 1936). It is convenient to consider at once a more general problem when the source S is shifted by a distance ζ with respect to the axis lens – observer, $L - O$ (see Fig. 5.1). For simplicity sake, we have approximated in this figure the real trajectory by a broken line. Since the deflection angle θ is small, the distance ξ coincides approximately with the impact parameter ρ. Then, recalling again that the angles θ and ϕ are small, we find the following relation for the true deflection:

$$\zeta = \xi \frac{l}{l_o} - l_s \theta = \xi \frac{l}{l_o} - l_s \frac{2r_g}{\xi} . \tag{5.5}$$

In the mentioned simplest case, when the source, lens, and observer are on

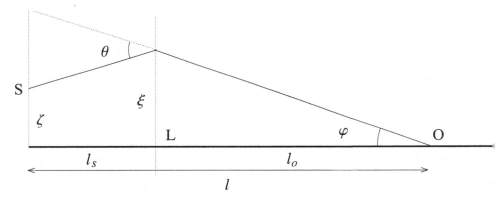

Fig. 5.1. Gravitational lens

the same axis, i.e. when $\zeta = 0$, we obtain from (5.5) that the fictitious radius of the ring, that is the image in the plane of the lens, is

$$\xi_0 = \sqrt{\frac{2r_g l_s l_o}{l}} ,$$

and its angular size equals

$$\phi = \frac{\xi_0}{l_o} = \sqrt{\frac{2r_g l_s}{l l_o}} .$$

Contrary to a possible naive dimensional estimate, this angle falls down not as the inverse characteristic distances themselves, but only as the square root of them. Still, the observation of the effect is practically impossible even if stars serve as both the source and the lens. However the effect gets observable when the source is a nebula, and the lens is a galaxy (*F. Zwicky*, 1937). Let us estimate the angular size of the ring for the case when this lens consists of 10^{10} stars with masses on the order of the Sun mass. Let the lens be situated at a distance on the order of 10^6 light years, or 10^{19} km, from us, and the distance to the source is much larger (i.e. $l \simeq l_s \gg l_o$). Then

$$\phi \sim \sqrt{\frac{6 \cdot 10^{10}}{10^{19}}} \sim 10^{-4} \text{ rad} \sim 10 \text{ angular seconds.}$$

Such a resolution is quite accessible for astronomers.

Let us address now a more general case when the lens does not lie on the axis source — observer. It is convenient here to go over to dimensionless variables

$$x = \frac{\xi}{\xi_0}, \quad y = \frac{\zeta}{\xi_0} \frac{l_o}{l}.$$

In these variables equation (5.5) reduces to

$$y = x - \frac{1}{x}, \tag{5.6}$$

with the obvious solution

$$x_\pm = \frac{1}{2} \left(y \pm \sqrt{y^2 + 4} \right).$$

Thus, in the general case, when the source S is shifted with respect to the direction to the lens L, the picture is different. Two images arise (see Fig. 5.2), one of them, I_1, is situated beyond the ring corresponding to the axisymmetric picture, another one, I_2, is inside the ring. The distance between them,

$$\Delta = x_+ - x_- = \sqrt{y^2 + 4},$$

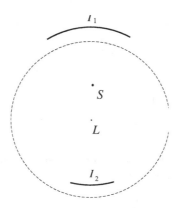

Fig. 5.2. Two images

is minimum for $y = 0$, i.e. for the axisymmetric position of the source, the lens and the observer. Since for such axisymmetric position both images should coalesce into a circle, it is clear that for $y \ll 1$ these images appear as arcs.

For the first time a gravitational lens was discovered in 1979. This was indeed a galaxy creating a double image of a quasar with the angular distance between its components of about 6 angular seconds. At present, few sources of radio waves are known which look like two arcs.

Problem

5.1. Consider a common optical lens that imitates the deflection of a ray of light by the gravitational field of a star. How does the thickness of such a lens change with its radius?

5.4 Microlenses

If the mass of an object, which acts as a lens, is not large, say, less than the mass of the Sun, to resolve the angle between the images is practically impossible. Nevertheless, even in this case the effect of gravitational lensing can be detected, due to the fact that when the images get closer, their total brightness increases. The brightness amplification K results from the growth of the total solid angle of the observed image as compared to the solid angle of the real source.

To estimate the effect, let us note that both ξ and ζ, as well as x and y, are in fact two-dimensional vectors that lie in the planes of the lens and the source, respectively. Evidently, the vector form of equation (5.6) is

$$\mathbf{y} = \mathbf{x} - \frac{\mathbf{x}}{x^2} . \tag{5.7}$$

Let us introduce coordinate axes in the planes of the lens and the source. We will label by the index 1 the axes, that lie in the plane passing through the source, lens, and observer, i.e. in the plane of Fig. 5.1; they are parallel to one another. We ascribe the index 2 to the axes orthogonal to the axes 1. The discussed ratio of the solid angles is obviously

$$K = \left(\frac{|\delta\xi_1 \delta\xi_2|}{l_o^2} \right) \left(\frac{|\delta\zeta_1 \delta\zeta_2|}{l^2} \right)^{-1} .$$

Here $\delta\xi_{1,2} (\delta\zeta_{1,2})$ are the sizes of the image (source) along the axes 1 and 2. In the dimensionless variables this ratio is

$$K = \frac{|\delta x_1 \delta x_2|}{|\delta y_1 \delta y_2|} = |\partial x_1 / \partial y_1| |\partial x_2 / \partial y_2| .$$

Both partial derivatives are taken at $y_2 = 0$. Therefore, in virtue of (5.7), x_1 and y_1 are related by the same equation (5.6):

$$y_1 = x_1 - \frac{1}{x_1},$$

and the relation between x_2 and y_2 is

$$y_2 = x_2 - \frac{x_2}{x_1^2}.$$

Thus, for the two different images the discussed relation of the solid angles is

$$K_{\pm} = \frac{\left(\sqrt{y^2 + 4} \pm y\right)^2}{4y\sqrt{y^2 + 4}}. \qquad (5.8)$$

For both images this ratio grows for small y:

$$K_{\pm} \simeq \frac{1}{2y}.$$

Therefore, the total brightness of the images increases as well:

$$K = K_+ + K_- \simeq \frac{1}{y}.$$

What happens when a star, acting as a gravitational lens, passes close to the line directed from the observer to the source? Even if one cannot resolve the arising double image, the observed brightness of the source grows as the lens approaches the line source — observer. This phenomenon, so-called microlensing, is of a rather special character: the increase of the brightness and its subsequent decrease are symmetric in time. Then, the brightness changes in the same way for all wave lengthes (the deflection angle (5.4) is independent of the wave length). And at last, since the phenomenon is extremely rare, it has one more distinctive feature: the repetition of the "flash" of a star caused by microlensing is practically excluded.

Not only the microlensing effect was detected. In this way a new class of celestial bodies was discovered — dwarf stars of low brightness, so-called brown dwarfs, which acted as microlenses.

Problem

5.2. Derive relation (5.8).

6

Variational Principle. Exact Solutions

6.1 Action for Gravitational Field. Energy-Momentum Tensor of Matter

The action S_g for the gravitational field should be an integral over the four-dimensional space, invariant under any coordinate transformations. It is natural to require that the field equations, resulting from variation of the action, should contain derivatives of the metric tensor $g_{\mu\nu}$ not higher than of second order. Then the integrand of S_g should contain derivatives of the metric not higher than of first order. In other words, it may depend on $g_{\mu\nu}$ and $\Gamma^\lambda_{\mu\nu}$ only. However, one cannot construct a scalar from these variables. Indeed, by going over into a locally inertial frame, one can make at any given point the metric flat and the Christoffel symbols equal to zero. However, in fact the scalar curvature R can serve as the integrand. Though it contains second derivatives of the metric, it depends linearly on them, so that one can get rid of these derivatives by means of integrating by parts.

Thus, let us demonstrate that the variation of the action $K \int d^4x \sqrt{-g} R$, with an appropriate choice of the constant K, results indeed in the Einstein equations. The variation of the integral gives

$$\delta \int d^4x \sqrt{-g}\, R = \delta \int d^4x \sqrt{-g}\, g^{\mu\nu} R_{\mu\nu}$$

$$= \int d^4x \sqrt{-g} \left(\delta g^{\mu\nu} R_{\mu\nu} + \frac{\delta\sqrt{-g}}{\sqrt{-g}} R + g^{\mu\nu} \delta R_{\mu\nu} \right). \qquad (6.1)$$

Then (see section 3.3),

$$\delta \sqrt{-g} = \frac{1}{2} \sqrt{-g}\, g^{\mu\nu} \delta g_{\mu\nu}\,.$$

Using the identity $g^{\mu\nu} g_{\mu\nu} = 4$, we present the second term in (6.1) as follows:

$$\frac{\delta\sqrt{-g}}{\sqrt{-g}} R = -\frac{1}{2}\, g_{\mu\nu} \delta g^{\mu\nu} R\,. \qquad (6.2)$$

Let us consider the last term of (6.1) in a locally inertial frame. Interchanging the operations of variation and differentiation, we find in this frame

$$\sqrt{-g}\,g^{\mu\nu}\delta R_{\mu\nu} = \sqrt{-g}\,g^{\mu\nu}\delta(\partial_\rho\Gamma^\rho_{\mu\nu} - \partial_\nu\Gamma^\rho_{\mu\rho}) = \sqrt{-g}\,\partial_\rho(g^{\mu\nu}\delta\Gamma^\rho_{\mu\nu} - g^{\mu\rho}\delta\Gamma^\nu_{\mu\nu}).$$

Of course, as follows from (3.19), the Christoffel symbol is no tensor since it transforms inhomogeneously under coordinate transformations. However, according to the same relation (3.19), the variation of the Christoffel symbol transforms homogeneously,

$$\delta\Gamma^\lambda_{\mu\nu} = \delta\Gamma'^\rho_{\sigma\tau}\frac{\partial x^\lambda}{\partial x'^\rho}\frac{\partial x'^\sigma}{\partial x^\mu}\frac{\partial x'^\tau}{\partial x^\nu},$$

and therefore is a tensor. Thus, the quantity

$$U^\rho = g^{\mu\nu}\delta\Gamma^\rho_{\mu\nu} - g^{\mu\rho}\delta\Gamma^\nu_{\mu\nu}$$

is a vector. Therefore, its divergence $\partial_\rho U^\rho$, which was written above in a locally inertial frame, can be rewritten in a generally covariant form:

$$U^\rho_{\;;\rho} = \frac{1}{\sqrt{-g}}\,\partial_\rho(\sqrt{-g}\,U^\rho).$$

In result, the last term in the variation of the action (6.1) reduces to the integral of a total divergence $\int d^4x\,\partial_\rho(\sqrt{-g}\,U^\rho)$ and hence can be omitted.

In this way we obtain the following variation of the gravitational action:

$$\delta S_g = K\,\delta\int d^4x\sqrt{-g}\,R = K\int d^4x\sqrt{-g}\left(R_{\mu\nu} - \frac{1}{2}g_{\mu\nu}R\right)\delta g^{\mu\nu}. \qquad (6.3)$$

To determine the constant K in it, we need the Einstein equation with the right-hand side, i.e. with the source. Therefore, let us find the variation of the action of matter by example, say, of a material point:

$$\delta S_m = -m\,\delta\int ds = -m\,\delta\int\sqrt{g_{\mu\nu}dx^\mu dx^\nu}$$

$$= -\frac{1}{2}m\int\frac{\delta g_{\mu\nu}dx^\mu dx^\nu}{\sqrt{g_{\rho\tau}dx^\rho dx^\tau}} = -\frac{1}{2}m\int ds\,u^\mu u^\nu\delta g_{\mu\nu}.$$

The last integral over ds transforms with the obvious identity

$$\int ds = \int d^4x\sqrt{-g}\,\frac{\delta(\mathbf{r} - \mathbf{r}(t))}{\sqrt{-g}u^0}, \qquad u^0 = \frac{dt}{ds},$$

to

$$-\frac{1}{2}\int d^4x\sqrt{-g}\,\rho u^\mu u^\nu\delta g_{\mu\nu} = -\frac{1}{2}\int d^4x\sqrt{-g}\,T^{\mu\nu}\delta g_{\mu\nu}.$$

Here

$$\rho(\mathbf{r}(t)) = m \, \frac{\delta(\mathbf{r} - \mathbf{r}(t))}{\sqrt{-g}u^0} \tag{6.4}$$

is the generally covariant mass density, and

$$T^{\mu\nu} = \rho u^\mu u^\nu \tag{6.5}$$

is the energy-momentum tensor of a point-like particle. At last, with the identity

$$T^{\mu\nu}\delta g_{\mu\nu} = T_{\rho\tau}g^{\rho\mu}g^{\tau\nu}\delta g_{\mu\nu} = -T_{\rho\tau}g^{\rho\mu}g_{\mu\nu}\delta g^{\tau\nu} = -T_{\mu\nu}\delta g^{\mu\nu},$$

we find

$$\delta S_m = \frac{1}{2} \int d^4x \sqrt{-g} \, T_{\mu\nu}\delta g^{\mu\nu}. \tag{6.6}$$

Now it follows from identities (6.3) and (6.6) that the variational principle

$$\delta(S_g + S_m) = 0$$

leads to the Einstein equation (4.1) (with vanishing cosmological constant) under the condition

$$K = -\frac{1}{16\pi k}.$$

Let us use relation (6.6) to derive the generally covariant expression for the energy-momentum tensor of electromagnetic field. The covariant action for this field is as follows:

$$S_{em} = \int d^4x \sqrt{-g} \, L_{em}; \quad L_{em} = -\frac{1}{16\pi} F_{\mu\nu}F^{\mu\nu} = -\frac{1}{16\pi} F_{\mu\rho}F_{\nu\tau}g^{\mu\nu}g^{\rho\tau}.$$

Variation with the account for relation (6.6) and formula (6.2) gives

$$T_{\mu\nu} = \frac{2}{\sqrt{-g}} \, \frac{\partial \sqrt{-g} \, L_{em}}{\partial g^{\mu\nu}} = -\frac{1}{4\pi} \left(F_{\mu\rho}F_{\nu\tau}g^{\rho\tau} - \frac{1}{4} g_{\mu\nu}F_{\rho\tau}F^{\rho\tau} \right). \tag{6.7}$$

Let us come back to the action

$$S_g = -\frac{1}{16\pi k} \int d^4x \sqrt{-g} \, R,$$

to exclude from it second derivatives. The terms with derivatives of Christoffel symbols in the integrand

$$\sqrt{-g} \, R = \sqrt{-g} \, g^{\mu\nu} R_{\mu\nu} = \sqrt{-g} \, g^{\mu\nu}(\partial_\rho \Gamma^\rho_{\mu\nu} - \partial_\nu \Gamma^\rho_{\mu\rho} + \Gamma^\rho_{\sigma\rho}\Gamma^\sigma_{\mu\nu} - \Gamma^\rho_{\sigma\nu}\Gamma^\sigma_{\mu\rho}),$$

after integrating by parts and omitting total derivatives, reduce to

$$-\Gamma^\sigma_{\mu\nu}\partial_\sigma(\sqrt{-g} \, g^{\mu\nu}) + \Gamma^\sigma_{\nu\sigma}\partial_\mu(\sqrt{-g} \, g^{\mu\nu}).$$

With the identities

$$\partial_\lambda \sqrt{-g} = \sqrt{-g}\, \Gamma^\rho_{\lambda\rho}, \quad \partial_\mu(\sqrt{-g}\, g^{\mu\nu}) = -\sqrt{-g}\, g^{\rho\tau}\, \Gamma^\nu_{\rho\tau},$$

$$g^{\mu\nu}_{\ ;\lambda} = g^{\mu\nu}_{\ ,\lambda} + \Gamma^\mu_{\lambda\tau} g^{\tau\nu} + \Gamma^\nu_{\lambda\tau} g^{\mu\tau} = 0\,,$$

the last expression reduces to

$$2\sqrt{-g}\, g^{\mu\nu}(\Gamma^\rho_{\mu\tau}\Gamma^\tau_{\nu\rho} - \Gamma^\rho_{\mu\nu}\Gamma^\tau_{\rho\tau})\,.$$

Thus, after eliminating second derivatives, the action for the gravitational field is as follows:

$$S_g = -\frac{1}{16\pi k} \int d^4x \sqrt{-g}\, g^{\mu\nu}(\Gamma^\rho_{\mu\tau}\Gamma^\tau_{\nu\rho} - \Gamma^\rho_{\mu\nu}\Gamma^\tau_{\rho\tau})\,. \tag{6.8}$$

Problem

6.1. Prove that the action S_g can be transformed also to

$$S_g = \frac{1}{32\pi k} \int d^4x \left[\Gamma^\rho_{\mu\nu}\, \partial_\rho(\sqrt{-g}\, g^{\mu\nu}) - \frac{1}{\sqrt{-g}}\, \partial_\mu\sqrt{-g}\, \partial_\nu(\sqrt{-g}\, g^{\mu\nu}) \right]\,.$$

6.2 Gravitational Field of Point-Like Mass

To solve the problem of the field of a point-like mass, we use the action in the form (6.8). We express the integrand through components of the metric possessing the spherical symmetry, and then obtain the field equations by the direct variation of the action with respect to the functions it depends on. In this way we do not need to calculate the Ricci tensor entering the Einstein equation (4.1).

As usual, we take the source for the origin of the reference frame. The spherical symmetry of our problem means that one can introduce the coordinates x_1, x_2, x_3 in such a way that ds^2 will go into itself under the transformations that look as Euclidean rotations of these coordinates. In this way we map the three-dimensional physical space onto the three-dimensional Euclidean one. Now the rotations in the physical space are mapped onto rotations in the Euclidean space that leave the quantity $r = \sqrt{x_1^2 + x_2^2 + x_3^2}$ invariant. In the Euclidean space there is no difference between co- and contravariant vectors, so that the use of the coordinates x_i with lower indices does not lead to confusion. In line with $r = \sqrt{\mathbf{x}^2}$, one can construct from $\mathbf{x}$ and $d\mathbf{x}$ two more scalars: $d\mathbf{x}^2$ and $\mathbf{x}d\mathbf{x}$. Therefore, in the static spherically symmetric case the interval can be written as

$$ds^2 = a^2(r)dt^2 - b(r)d\mathbf{x}^2 - c(r)(\mathbf{x}d\mathbf{x})^2\,. \tag{6.9}$$

With the change of variables of the type $\mathbf{x} \to f(r)\mathbf{x}$, one can make $b(r) = 1$. Then the spherically symmetric metric is expressed through two unknown functions of r:

$$ds^2 = a^2(r)dt^2 - d\mathbf{x}^2 - c(r)(\mathbf{x}d\mathbf{x})^2,$$

$$g_{00} = a^2(r), \quad g_{mn} = -\delta_{mn} - c(r)x_m x_n. \tag{6.10}$$

With this choice of the reference frame, the space metric

$$dl^2 = d\mathbf{x}^2 + c(r)(\mathbf{x}d\mathbf{x})^2 = dr^2 + r^2(d\theta^2 + \sin^2\theta\, d\phi^2) + cr^2 dr^2$$

$$= d^2(r)dr^2 + r^2(d\theta^2 + \sin^2\theta\, d\phi^2), \quad d^2(r) = 1 + c(r)r^2,$$

is such that the infinitesimal arc of a circle in the plane $\theta = \pi/2$ is $dl = rd\phi$, i.e. the length of a circle, with its center in the origin, is $2\pi r$ as usual. Simple calculations demonstrate that in metric (6.10) only the following components of the Christoffel symbol $\Gamma_{\rho,\mu\nu}$ do not vanish:

$$\Gamma_{0,i0} = -\Gamma_{i,00} = aa'\frac{x_i}{r}, \quad \Gamma_{i,jk} = -\left(cx_i\delta_{jk} + \frac{1}{2}\frac{c'}{r}x_i x_j x_k\right).$$

Obviously, $g^{00} = 1/a^2$, so that

$$\Gamma^0_{i0} = \frac{a'}{a}\frac{x_i}{r}.$$

To find Γ^i_{00} and Γ^i_{jk}, let us consider the expression $g^{km}g_{mn}x_n$. On the one hand, due to the identity $g^{km}g_{mn} = \delta^k_n$, it equals x_k. On the other hand, a direct calculation gives $g_{mn}x_n = -(\delta_{mn} + cx_m x_n)x_n = -d^2 x_m$, so that $g^{km}g_{mn}x_n = -d^2 g^{km}x_m$. It is clear now that

$$g^{km}x_m = -\frac{1}{d^2}x_k.$$

Thus we find easily

$$\Gamma^i_{00} = \frac{aa'}{d^2}\frac{x_i}{r}, \quad \Gamma^i_{jk} = \frac{x_i}{d^2}\left(c\delta_{jk} + \frac{1}{2}\frac{c'}{r}x_j x_k\right).$$

At last, let us note that due to the spherical symmetry of the problem, it is sufficient to calculate the integrand of the action at a single point, $x_1 = r$, $x_2 = x_3 = 0$. Only the following components of the space metric and the Christoffel symbols are nonvanishing in this point:

$$g_{11} = -d^2, \quad g_{22} = g_{33} = -1, \quad g^{11} = -\frac{1}{d^2}, \quad g^{22} = g^{33} = -1,$$

$$\Gamma^0_{10} = \Gamma^0_{01} = \frac{a'}{a}, \quad \Gamma^1_{00} = \frac{aa'}{d^2}, \quad \Gamma^1_{22} = \Gamma^1_{33} = \frac{r}{d^2}c,$$

$$\Gamma^1_{11} = \frac{r}{d^2}\left(c + \frac{1}{2}c'r\right) = \frac{(d^2)'}{2d^2} = \frac{d'}{d}.$$

When substituting these expressions into formula (6.8), the terms

$$g^{00}(\Gamma^\rho_{0\tau}\Gamma^\tau_{0\rho} - \Gamma^\rho_{00}\Gamma^\tau_{\rho\tau}) \quad \text{and} \quad g^{11}(\Gamma^\rho_{1\tau}\Gamma^\tau_{1\rho} - \Gamma^\rho_{11}\Gamma^\tau_{\rho\tau})$$

cancel, and other terms produce

$$S_g = -\frac{1}{8\pi k} \int d^4x(ad)' \frac{cr}{d^2} = -\frac{1}{2k} \int dtdr(ad)'r\left(1 - \frac{1}{d^2}\right).$$

It is convenient now to introduce new independent functions $u = r\left(1 - 1/d^2\right)$, $w = ad$. Then the variation of the action

$$S_g = -\frac{1}{2k} \int dtdr u w'$$

is trivial and gives $w = c_1$, $u = c_2$. Coming back to the old functions, we find easily

$$d^2 = \left(1 - \frac{c_2}{r}\right)^{-1}, \quad a^2 = c_1^2\left(1 - \frac{c_2}{r}\right).$$

Since a^2 enters the interval ds^2 only through a^2dt^2, by changing the scale of time we can put $c_1 = 1$. And finally, recalling that at large distances from a gravitating mass M $g_{00} = 1 - 2kM/r$, we obtain $c_2 = 2kM = r_g$. In this way we arrive at the metric for a gravitating point-like mass (K. Schwarzschild, 1916):

$$ds^2 = \left(1 - \frac{2kM}{r}\right) dt^2 - \left(1 - \frac{2kM}{r}\right)^{-1} dr^2 - r^2(d\theta^2 + \sin^2\theta d\phi^2). \quad (6.11)$$

Problems

6.2. Find the surface of rotation on which the geometry is the same as that on the "plane" passing through the origin in the Schwarzschild solution.

6.3. Find the spherically symmetric solution of the Einstein equations with the cosmological constant. Estimate the upper limit on the value of this constant, following from the fact that for Pluto (the radius of the orbit of this planet is $\sim 10^{15}$ cm) the Kepler laws are valid with an accuracy better than 10^{-5}. Formulate this upper limit for the corresponding effective mass density τ^{00} (see section 4.1).

6.3 Harmonic and Isotropic Coordinates. Relativistic Correction to the Newton Law

Let us note now that the space part of the metric (6.11) does not go over for $r \to \infty$ into the solution $g_{mn} = -\delta_{mn}(1 + 2kM/r)$, obtained in section 4.2 for the case of a weak field. The reason is that metric (6.11) and the above weak-field solution correspond to different choices of the radial coordinate. The simplest way to reproduce that weak-field limit, starting with metric (6.11), is to shift in (6.11) the radial coordinate as follows:

$$r \to r + kM. \tag{6.12}$$

With this shift we arrive at the interval

$$ds^2 = \frac{1 - kM/r}{1 + kM/r} \, dt^2 - \frac{1 + kM/r}{1 - kM/r} \, dr^2 - (1 + kM/r)^2 (d\theta^2 + \sin^2\theta d\phi^2), \tag{6.13}$$

or

$$ds^2 = \frac{1 - kM/r}{1 + kM/r} \, dt^2 - (1 + kM/r)^2 \, d\mathbf{r}^2$$
$$- \left(\frac{kM}{r} \right)^2 \frac{1 + kM/r}{1 - kM/r} \left(\frac{\mathbf{r} d\mathbf{r}}{r} \right)^2. \tag{6.14}$$

The metric

$$g_{00} = \frac{1 - kM/r}{1 + kM/r}, \quad g_{0n} = 0,$$

$$g_{mn} = -(1 + kM/r)^2 \, \delta_{mn} - \left(\frac{kM}{r} \right)^2 \frac{1 + kM/r}{1 - kM/r} \frac{r_m r_n}{r^2} \tag{6.15}$$

of interval (6.14) (or (6.13)) not only agrees with the linear harmonic gauge $\partial_\mu h_{\mu\nu} - \frac{1}{2}\partial_\nu h_{\mu\mu} = 0$ of section 4.2. It satisfies a more general harmonic condition

$$\partial_\mu \left(\sqrt{-g} g^{\mu\nu} \right) = 0, \tag{6.16}$$

that is not confined to the weak-field approximation. The coordinates satisfying condition (6.16) are called harmonic.

On the other hand, by substitution

$$r = \rho \left(1 + \frac{r_g}{4\rho} \right)^2, \tag{6.17}$$

we obtain from (6.11) such an expression for the interval, where the space metric is isotropic. It differs from the Euclidean space metric by an overall factor only (i.e. is conformally Euclidean):

$$ds^2 = \left(\frac{1 - r_g/4\rho}{1 + r_g/4\rho} \right)^2 dt^2 - \left(1 + \frac{r_g}{4\rho} \right)^4 [d\rho^2 + \rho^2(d\theta^2 + \sin^2\theta \, d\phi^2)]. \tag{6.18}$$

Obviously, the asymptotics of this metric for $r \gg r_g$ also coincides with that found in section 4.2.

Let us find now the relativistic correction to the gravitational interaction of two bodies with comparable masses m_1 and m_2. Dimensional arguments (recall that km/c^2 has the dimension of length) combined with the requirement of symmetry under permutation $m_1 \leftrightarrow m_2$, dictate that the corresponding velocity-independent correction to the Newton law should have the structure

$$a \, \frac{k^2 m_1 m_2 (m_1 + m_2)}{c^2 r^2} \, .$$

To find the dimensionless numerical constant a in this expression, we expand the Lagrangian for a light particle of mass m_1 in the gravitational field of a heavy body with mass m_2 in harmonic coordinates to first order in $1/c^2$:

$$L = -m_1 \sqrt{g_{\mu\nu} u^\mu u^\nu} = -m_1 c^2 + \frac{m_1 v^2}{2} + \frac{k m_1 m_2}{r} - \frac{1}{2} \frac{k^2 m_1^2 m_2}{c^2 r^2}$$

$$+ \frac{3}{2} \frac{k m_1 m_2 v^2}{c^2 r} \, . \tag{6.19}$$

Thus, in the case of a heavy mass m_2, the static gravitational potential

$$U^{(0)}(r) = -\frac{k m_1 m_2}{r}$$

acquires the relativistic correction $k^2 m_2^2 m_1 / 2 c^2 r^2$.

For comparable masses m_1 and m_2, restoring the symmetry between m_1 and m_2, we arrive at the relativistic correction (*A. Einstein, L. Infeld, B. Hoffmann*, 1938; *A. Eddington, G. Clark*, 1938)

$$U^{(2)}(r) = \frac{1}{2} \frac{k^2 m_1 m_2 (m_1 + m_2)}{c^2 r^2} \, . \tag{6.20}$$

For the derivation of correction (6.20), it was rather crucial to use the harmonic coordinates satisfying subsidiary condition (6.16) since this condition does not violate the required symmetry under $m_1 \leftrightarrow m_2$. As to the Schwarzschild coordinates, with their origin chosen at one of the particles, they are not appropriate for this problem.

Let us note, however, that one can arrive at correction (6.20) starting with the isotropic coordinates.

Problems

6.4. Prove that metric (6.14) satisfies the harmonic condition (6.16).

6.5. Derive transformation (6.17) that changes the Schwarzschild coordinates into the isotropic ones.

6.4 Precession of Orbits in the Schwarzschild Field

A simple-minded dimensional estimate for the relative magnitude of the precession is again $\sim r_g/r$, where r is the characteristic radius of the orbit. In other words, during one unperturbed turn of the radius-vector (by the angle 2π) the semi-axis of the elliptic orbit precesses by the angle

$$\delta\phi \sim \frac{2\pi r_g}{r}\,. \tag{6.21}$$

We start the quantitative consideration of the particle motion with the equation connecting its energy $E = p_0$ with the three-dimensional momentum $\mathbf{p}$ and mass m:

$$g^{\mu\nu}p_\mu p_\nu - m^2 = 0\,. \tag{6.22}$$

For the solution of the problem, it is convenient to use the isotropic coordinates (6.18). For a diagonal metric its contravariant components $g^{\mu\nu}$ are inverse to the covariant ones, so that the explicit form of equation (6.22) is here as follows:

$$\left(\frac{1+r_g/4\rho}{1-r_g/4\rho}\right)^2 E^2 - \left(1 + \frac{r_g}{4\rho}\right)^{-4}\left(p_\rho^2 + \frac{L^2}{r^2}\right) - m^2 = 0\,. \tag{6.23}$$

The motion of a particle in a centrally symmetric gravitational field, as well as in any other central field, takes place in a plane passing through the origin. We choose for this plane the plane $\theta = \pi/2$. The energy E and the orbital angular momentum L are integrals of motion.

Here we go beyond the linear approximation and include terms of second order in r_g/ρ. Multiplying equation (6.23) by $(1 + r_g/(4\rho))^4$ and expanding thus obtained coefficients in r_g/ρ, we get

$$\left(1 + 2\frac{r_g}{\rho} + \frac{15}{8}\frac{r_g^2}{\rho^2}\right)E^2 - \left(1 + \frac{r_g}{\rho} + \frac{3}{8}\frac{r_g^2}{\rho^2}\right)m^2 - \left(p_\rho^2 + \frac{L^2}{r^2}\right) = 0\,. \tag{6.24}$$

Now we put $E = m + \varepsilon$, where ε is the nonrelativistic integral of energy, and keep the terms not higher than second order in $1/c$ (c is the velocity of light, here we do not write it down explicitly). In the relation, arising in this way,

$$2m\varepsilon + \varepsilon^2 + (m^2 + 4m\varepsilon)\frac{r_g}{\rho} + \frac{3}{2}\frac{r_g^2 m^2}{\rho^2} - \left(p_\rho^2 + \frac{L^2}{r^2}\right) = 0\,,$$

one can drop ε^2 as compared to $2m\varepsilon$, and $4m\varepsilon$ as compared to m^2 in the factor at r_g/ρ. Obviously, neither of these corrections contributes to the precession of the orbit. The resulting expression can be rewritten as

$$\varepsilon = \frac{1}{2m}\left(p_\rho^2 + \frac{L^2}{\rho^2}\right) - \frac{kmM}{\rho} - \frac{3}{4}\frac{mr_g^2}{\rho^2}\,.$$

Thus, the problem is reduced to the motion in the Newton potential with the perturbation

$$\delta U(\rho) = -\frac{3k^2 m M^2}{\rho^2} .$$

Just the same result follows from the solution of the problem in the Schwarzschild metric (6.11). A simple calculation[1] demonstrates that this perturbation results in the rotation of the semi-major axis a by the angle

$$\delta \phi = \frac{3\pi r_g}{a(1-e^2)} = \frac{6\pi k M}{a(1-e^2)} \tag{6.25}$$

during one turn. Here e is the eccentricity of the unperturbed elliptic orbit. Our initial estimate (6.21) is confirmed (up to a factor 3/2).

As to the planets of our solar system, the maximum effect should be expected for Mercury, since the radius of its orbit is the smallest one. However, even for it the effect is tiny: formula (6.25) gives for the shift of the Mercury perihelion only $43.0''$ per century. Nevertheless such an anomaly in the Mercury motion on the level of $45'' \pm 5''$ per century, incomprehensible at that time, had been known to astronomers before Einstein. Its natural explanation was the first triumph of GR. For a long time the rotation of the Mercury perihelion was the only really observed nonlinear effect of GR. At present this prediction of GR is confirmed by radar measurements with an accuracy of about 1%.

Below we present the predictions of GR (first number) and the results of measurements (second number) for Mercury and other objects. The units are the same: angular seconds per century.

Mercury: 43.03 , 43.11 ± 0.45 .

Venus: 8.6 , 8.4 ± 4.8 .

Earth: 3.8 , 5.0 ± 1.2 .

Icarus: 10.3 , 9.8 ± 0.8 .

Large eccentricity of the orbit of the asteroid Icarus enhances the effect (see formula (6.25)), and at the same time allows one to measure the effect with better accuracy.

One may expect that the effect will be much more pronounced in the motion of binary stars, since the gravitational fields in these systems are much stronger. Indeed, careful investigations of the binary pulsar B1913+16 (B means binary pulsar, numbers refer to the coordinates on the celestial sphere:

---

[1]See, for instance, L.D. Landau and E.M. Lifshitz, *Mechanics*, §15, Problem 3.

the direct ascension is $19^h 13^m$, the inclination is $16°$) have shown that in this binary the orbit periastron rotates by $4.2°$ per year. By the way, in such a way the masses of the binary components were measured with high accuracy: 1.4414 ± 0.0002 and 1.3867 ± 0.0002 solar masses, respectively. It is no wonder that the periastron rotation is so large here: though the masses of the components are quite comparable to the solar mass, the distance between them, 1.8×10^6 km, is small as compared, say, to the radius of the Mercury orbit, 0.6×10^8 km.

Problems

6.6. Find the orbit precession, due to the relativistic correction, in the attracting Coulomb potential.

6.7. Find the orbit precession, due to the relativistic correction, in the attracting scalar potential, assuming that this potential is introduced into the equation $p_\mu p_\mu = m^2$ by means of the substitution $m \to m + \phi$.

6.5 Retardation of Light in the Field of the Sun

The effect discussed in the present section is linear in r_g, and from this point of view should be considered in the previous chapter. However this effect is of interest not only in relation to the experimental check of GR. Its detailed consideration is quite instructive in the sense of comparison of the Schwarzschild and harmonic coordinates. Due to it, this section is included in the present chapter.

So, let us consider the propagation of a signal from the point E, $\mathbf{r}_1 = (x_1, y)$, to the point V, $\mathbf{r}_2 = (x_2, y)$, in the gravitational field created by a mass M, situated at the point S, $\mathbf{r}_0 = 0$, (see Fig. 6.1). We mean in fact the influence of the gravitational field of the Sun on the propagation of a radar signal sent from Earth to Venus. Hence the notation of the points in Fig. 6.1.

At first we solve the problem in the Schwarzschild coordinates. To this end the interval (6.11) is rewritten as follows:

$$ds^2 = \left(1 - \frac{r_g}{r}\right) dt^2 - dr^2 - \frac{r_g}{r}\left(1 - \frac{r_g}{r}\right)^{-1} dr^2 = 0.$$

With the identity $dr = (\mathbf{r} \cdot d\mathbf{r})/r$, we obtain to first order in r_g

$$dt = dx \left(1 + \frac{r_g}{2r} + \frac{r_g x^2}{2r^3}\right).$$

The total transit time is

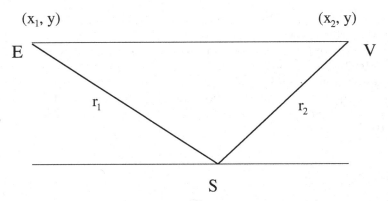

Fig. 6.1. Radar signal from Earth to Venus

$$T = x_2 - x_1 + r_g \ln \frac{x_2 + r_2}{x_1 + r_1} - \frac{r_g}{2} \left(\frac{x_2}{r_2} - \frac{x_1}{r_1} \right) \qquad (6.26)$$

(for the location of the planets as in Fig. 6.1, $x_1 < 0$). Obviously, the retardation of the signal ΔT is described by the terms proportional to r_g in this expression.

In the harmonic coordinates (labeled now with primes to distinguish them from the Schwarzschild ones) we have, correspondingly,

$$ds^2 = \frac{1 - kM/r'}{1 + kM/r'} \, dt^2 - (1 + kM/r')^2 \, dx'^2 - \left(\frac{kM}{r'} \right)^2 \frac{1 + kM/r'}{1 - kM/r'} \, dx'^2 = 0 \, ,$$

and obtain, again to first order in r_g,

$$dt = dx' \left(1 + \frac{r_g}{r'} \right) .$$

Now the total transit time is

$$T' = x_2' - x_1' + r_g \ln \frac{x_2 + r_2}{x_1 + r_1} \, . \qquad (6.27)$$

Since we confine to effects of first order in r_g, in the logarithmic term here the difference between the harmonic and Schwarzschild coordinates is neglected, i.e. primes in this term are omitted.

No wonder that in different coordinates, Schwarzschild and harmonic ones, the results for retardation,

$$\Delta T = r_g \ln \frac{x_2 + r_2}{x_1 + r_1} - \frac{r_g}{2} \left(\frac{x_2}{r_2} - \frac{x_1}{r_1} \right) \quad \text{and} \quad \Delta T' = r_g \ln \frac{x_2 + r_2}{x_1 + r_1} \, ,$$

respectively, are also different. Indeed, one can obtain formula (6.27) directly from (6.26) with the change of variables (6.12). Under it, the nonlogarithmic term in ΔT is cancelled by a correction $\sim r_g$ arising in $x_2 - x_1$.

In other words, formulae (6.26) and (6.27) differ since r and r' therein correspond to different physical distances (see (6.12)). For instance, let us consider two circular orbits, such that the numerical value of radius r for one of them is equal to the numerical value of r' for another. These orbits have in particular different periods, and the latter are directly observable.[2]

Still, the natural question arises: how should one compare the theory with experiment? The answer is as follows. There is no way to measure $x_2 - x_1$ directly since measuring rods are of no use, and light signals do not differ from the radar ones. But $x_2 - x_1$ can be expressed in terms of r_1, r_2, and ϕ (see Fig. 6.1). On the other hand, neither of the last three parameters can be directly measured with required accuracy. However, r_1 and r_2 can be expressed via the observable orbital periods, eccentricities, and times elapsed since the perihelions. To determine the angle ϕ one needs also the time elapsed since conjunction of the planets. It goes without saying that the predictions for the experiment, obtained in this way from formulae (6.26) and (6.27), are identical.

The results of measurements of the signal retardation in the gravitational field of the Sun agree for Venus with the prediction of GR within their accuracy that constitutes 3 to 4%. The best experiments performed with satellites with the active reflection confirm this result of GR with the accuracy of 0.2%.

Problems

6.8. Derive formula (6.27) from (6.26) with the change of variables (6.12).

6.9. Prove that the third Kepler law is valid for circular orbits in the Schwarzschild coordinates, but not in the isotropic ones.

6.10. Estimate the correction to the retardation time due to the signal deflection by the Sun.

6.11. A particle has an initial velocity v_0 at infinity and falls radially to a black hole. How does its velocity change with the distance? Under the assumption of a weak gravitational field, find the value of v_0 for which the particle velocity remains constant. (*M. Carmeli*, 1972; *S.I. Blinnikov, M.I. Vysotsky, L.B. Okun'*, 2001).

---

[2]One cannot but recall here the well-known comment by V.A. Fock: "Physics is essentially a simple science. The main problem in it is to understand which letter means what."

6.6 Motion in Strong Gravitational Field

Let us consider now the motion of a point-like particle in a strong gravitational field. The problem is solved conveniently with the Hamilton-Jacobi equation

$$g^{\mu\nu}\partial_\mu S\,\partial_\nu S - m^2 = 0\,.$$

For the motion in the plane $\theta = \pi/2$ this equation appears in the Schwarzschild coordinates as follows:

$$\left(1 - \frac{r_g}{r}\right)^{-1}(\partial_t S)^2 - \left(1 - \frac{r_g}{r}\right)(\partial_r S)^2 - \frac{1}{r^2}(\partial_\phi S)^2 - m^2 = 0\,. \qquad (6.28)$$

Its solution can be presented in the form

$$S = -Et + L\phi + s(r)\,.$$

We are interested here in the radial motion of the particle, when its orbital angular momentum vanishes, $L = 0$. Then

$$s'(r) = -\left(1 - \frac{r_g}{r}\right)^{-1}\sqrt{E^2 - m^2\left(1 - \frac{r_g}{r}\right)}\,.$$

The dependence $r = r(t)$ is found with the usual equation $\partial S/\partial E = \text{const}$:

$$t - t_0 = -\int_{r_0}^{r}\frac{dr}{(1 - r_g/r)\sqrt{1 - (1 - r_g/r)\,m^2/E^2}}\,.$$

Our choice of the sign for the radical corresponds to the motion of the particle to the center, r decreases with the increase of t. As the initial condition for $t = 0$ we choose $r = r_0$, $\dot{r} = 0$. Now,

$$\left(1 - \frac{r_g}{r_0}\right)^{-1}E^2 = m^2\,, \quad \text{or} \quad \frac{m^2}{E^2} = \left(1 - \frac{r_g}{r_0}\right)^{-1}\,.$$

For simplicity sake, assume also that $r_0 \gg r_g$. Then we obtain

$$t = -\int_{r_0}^{r}\frac{dr\sqrt{r^3/r_g}}{r - r_g}\,.$$

For $r \to r_g$ we find from it

$$t \simeq -\int^{r}\frac{dr\,r_g}{r - r_g} \simeq -r_g\ln\frac{r - r_g}{r_g}\,, \quad \text{or} \quad r - r_g \simeq r_g e^{-t/r_g}\,.$$

Thus, from the point of view of a distant observer, the particle approaches the gravitational radius asymptotically, reaching it only for $t \to \infty$. In the course of the approach the particle velocity dr/dt tends asymptotically to zero. In the last chapter of the book we will come back to this problem.

Let us consider now the radial propagation of light from a point r to a point $r_0 > r$. Here $ds^2 = 0$, so that $dt = dr\sqrt{|g_{rr}|}/\sqrt{g_{00}}$, and the time of light propagation,

$$\Delta t = \int_r^{r_0} dr \left(1 - \frac{r_g}{r}\right)^{-1} = r_0 - r + r_g \ln \frac{r_0 - r_g}{r - r_g}, \qquad (6.29)$$

tends to infinity with the initial point r approaching r_g. The signal from the surface $r = r_g$ travels for infinite time. Moreover, the frequency of light as observed by a distant observer, also decreases when the source approaches r_g, changing according to

$$\omega \sim 1 - \frac{r_g}{r}. \qquad (6.30)$$

One factor $\sqrt{1 - r_g/r}$ in this relation arises as usual from $\sqrt{g_{00}}$, and the second one from the Doppler effect due to the motion of the source to the center.

However the fall of a particle to the center looks absolutely different for an observer freely falling together with this particle. The interval of its proper time is

$$d\tau = \sqrt{g_{00}dt^2 + g_{rr}dr^2} = \sqrt{g_{00}(dt/dr)^2 + g_{rr}}\, dr = \left(\frac{r_g}{r} - \frac{r_g}{r_0}\right)^{-1/2} dr.$$

Clearly, the particle reaches the Schwarzschild sphere during finite proper time

$$\tau = - \int_{r_0}^r \left(\frac{r_g}{r} - \frac{r_g}{r_0}\right)^{-1/2} dr.$$

By the way, near the gravitational radius the velocity of this particle, according to its proper time, tends to c.

After the particle crosses the Schwarzschild sphere, it moves to the center, $r = 0$, and reaches it also during finite time. Here, for $r < r_g$, g_{00} becomes negative, and g_{rr} becomes positive. In other words, inside the Schwarzschild sphere t becomes a space-like coordinate, and r becomes a time-like one! The motion of a particle for $r < r_g$ shows how the "time" r flows in this region: it flows to the origin $r = 0$. But it means that even if one would try to reverse the direction of motion in the region $r < r_g$, say, by switching on a powerful rocket, the attempt would fail, regardless of how powerful the rocket is. Inside the sphere $r = r_g$ the motion is possible to the center only.

Thus, the Schwarzschild sphere is the horizon of events, a one-way gate, it does not let out to a remote observer any signal. Hence the name of such an object — black hole.[3]

It means in particular that the reference frame of a remote observer, inertial at infinity, is incomplete: it does not describe the motion inside the sphere

---

[3]We will see in the last chapter that this name is not quite accurate.

$r = r_g$. Besides, in it g_{00} turns to 0 and g_{rr} turns to infinity at $r = r_g$. However, this singularity is special for the Schwarzschild system of coordinates. The invariants of the metric are regular on the surface $r = r_g$. This is obvious for the determinant of the metric tensor, $g = -r^4 \sin^2 \theta$, and can be proven by a direct calculation for the invariant $R_{\mu\nu\rho\tau} R^{\mu\nu\rho\tau}$. However, the last invariant turns to infinity at $r = 0$. At this point the metric has a true singularity.

To construct a reference frame free of the singularity at $r = r_g$, one can take a set of freely falling particles of dust, enumerate them with radial marks, and choose the proper time of a particle as the time coordinate (G. Lemaitre, 1938). Indeed, there is no singularity in this reference frame. But neither there are particles at rest inside the horizon. By the way, in this comoving reference frame not only the invariant $R_{\mu\nu\rho\tau} R^{\mu\nu\rho\tau}$ remains finite at $r = r_g$, but all components of the Riemann tensor are finite as well. In other words, in this frame the tidal forces acting upon an extended, non-pointlike body are finite (see section 3.7).

Problems

6.12. Find the radii of circular orbits in the field of a black hole (S.A. Kaplan, 1949).

6.13. Find the cross-section of the gravitational capture by a black hole of a nonrelativistic (at infinity) particle, and the correction of first order in v/c to this cross-section (Ya.B. Zel'dovich, I.D. Novikov, 1964).

6.14. Find the cross-section of the gravitational capture by a black hole of an ultrarelativistic (at infinity) particle, and the correction of first order in $1/\gamma$ to this cross-section (Ya.B. Zel'dovich, I.D. Novikov, 1964).

6.15. A particle with the velocity $v_\infty \ll 1$ at infinity and with the impact parameter $\rho = 2r_g(1 + \delta)/v_\infty$, $\delta \ll 1$, is scattered by a black hole and goes again to infinity. Describe qualitatively the motion of this particle (Ya.B. Zel'dovich, I.D. Novikov, 1964). What is its velocity near the black hole?

6.16. Ultrarelativistic particle with the impact parameter

$$\rho = (3\sqrt{3}/2)\, r_g(1 + \delta), \quad \delta \ll 1,$$

is scattered by a black hole and goes again to infinity. Describe qualitatively the motion of this particle (Ya.B. Zel'dovich, I.D. Novikov, 1964). What is its velocity near the black hole?

6.17. A black hole (serving as a gravitational lens), a point source of light, and an observer are perfectly aligned, just in this order. Describe qualitatively the picture seen by the observer around the black hole (D.E. Holz, J.A. Wheeler, 2002).

6.18. Derive relation (6.22).

6.7 Gravitational Field of Charged Point-Like Mass

Since any initially charged astrophysical object would be neutralized rapidly by the surrounding matter, the case of a charged star is unrealistic by itself. However, the considered problem is undoubtedly of a methodological interest as a sufficiently simple, but nontrivial generalization of the Schwarzschild solution.

Even if a point-like source is charged, its metric still has the structure (6.10). To find in this case the functions $a(r)$ and $c(r)$, let us consider at first the field of the charge. Obviously, it has no magnetic field, as well as in the case when there is no gravity. To find the electric field, we use the covariant Maxwell equation:

$$F^{\mu\nu}{}_{;\mu} = 4\pi j^{\nu}. \tag{6.31}$$

The left-hand side of its zeroth component has the following explicit form:

$$F^{\mu 0}{}_{;\mu} = \frac{1}{\sqrt{-g}} \partial_m (\sqrt{-g}\, F^{m0}).$$

As to the right-hand side of this component, the invariant charge density therein is

$$\rho_e(\mathbf{r}) = e\, \frac{\delta(\mathbf{r} - \mathbf{r}(t))}{\sqrt{-g}\, u^0}, \tag{6.32}$$

just as the invariant mass density is given by formula (6.4). Correspondingly,

$$j^0 = \rho_e(\mathbf{r}) u^0 = e\, \frac{\delta(\mathbf{r} - \mathbf{r}(t))}{\sqrt{-g}}.$$

Arising in this way, equation

$$\partial_m (\sqrt{-g}\, F^{m0}) = 4\pi e \delta(\mathbf{r} - \mathbf{r}(t))$$

is solved with the Gauss theorem immediately:

$$\sqrt{-g}\, F^{r0} = \frac{e}{r^2}.$$

Hence the radial electric field is

$$F_{0r} = -F_{r0} = -\frac{g_{00} g_{rr}}{\sqrt{-g}}\, \frac{e}{r^2} = ad\, \frac{e}{r^2}. \tag{6.33}$$

The action for the electromagnetic field is in this case as follows:

$$S_{em} = -\frac{1}{8\pi} \int d^4 x \sqrt{-g}\, F^{0r} F_{0r} = \frac{1}{2} \int dt dr r^2\, \frac{1}{ad}\, F_{0r}^2.$$

Now, the total action is (in the same variables $u = r(1 - 1/d^2)$ and $w = ad$)

$$S = S_g + S_{em} = -\frac{1}{2} \int dt dr \left(\frac{1}{k} uw' - \frac{r^2 F_{0r}^2}{w} \right). \tag{6.34}$$

The variation of this action with respect to the metric should be performed at fixed covariant field components, F_{0r} in the present case, since just for them the definition in curvilinear coordinates looks the same as in cartesian ones: $F_{\mu\nu} = \partial_\mu A_\nu - \partial_\nu A_\mu$; it contains neither metric, nor Christoffel symbols. Let us note also that the variation of the total action (including $-e \int A_\mu dx^\mu$ in line with S_g and S_{em}) just with respect to the covariant components A_μ results in the Maxwell equation (6.31) in the Riemann space.

The variation of the obtained action (6.34) with respect to u gives $w' = 0$, $w = c_1$. As well as in the case of the Schwarzschild solution, we put $c_1 = 1$, i.e. $w = ad = 1$. Then the variation with respect to w results in $u' = kr^2 F_{0r}^2/w^2 = ke^2/r^2$, or $u = r(1 - 1/d^2) = c_2 - ke^2/r$. It follows now that

$$a^2 = d^{-2} = 1 - \frac{c_2}{r} + \frac{ke^2}{r^2}.$$

Recalling again that at large distances from the gravitating mass

$$g_{00} = 1 - \frac{2kM}{r},$$

we obtain $c_2 = 2kM = r_g$. In this way we arrive at the metric created by a charged point-like mass (*H. Reissner*, 1916; *G. Nordtröm*, 1918):

$$ds^2 = \left(1 - \frac{2kM}{r} + \frac{ke^2}{r^2}\right) dt^2$$

$$- \left(1 - \frac{2kM}{r} + \frac{ke^2}{r^2}\right)^{-1} dr^2 - r^2(d\theta^2 + \sin^2\theta d\phi^2). \qquad (6.35)$$

The horizon radius here is the root

$$r_{rn} = kM + \sqrt{k^2M^2 - ke^2} \qquad (6.36)$$

of equation

$$1 - \frac{2kM}{r} + \frac{ke^2}{r^2} = 0.$$

Of course, of the two roots of this equation we have chosen that one which goes over into $r_g = 2kM$ for $e = 0$. The Reissner – Nordström solution has a physical meaning only for $e^2 \leq kM^2$. The charged black hole with $e^2 = kM^2$ is called extremal.

It is useful to present another derivation of the Reissner – Nordström solution, a less rigorous one, but one that demonstrates explicitly the origin of the term ke^2/r^2 in (6.35). Let us start with the Schwarzschild solution (6.11). When in line with the point-like mass M_0, there is a distributed mass $m(r)$, it is natural to perform in expression (6.11) the substitution

$$M \to M_0 + m(r).$$

In the present case $m(r)$ is nothing but the part of the electrostatic energy of the charge e that is confined inside the sphere of the radius r:

$$m(r) = 4\pi \int dr \, r^2 \frac{F_{0r}^2}{8\pi} = \frac{e^2}{2} \int_{r_0}^{r} \frac{dr}{r^2} = \frac{e^2}{2} \left(\frac{1}{r_0} - \frac{1}{r} \right).$$

As usual, the electrostatic energy of a classical point-like charge diverges linearly, and to obtain a finite result one has to introduce a minimum distance r_0. The term $e^2/(2r_0)$, arising in this way, corresponds to the classical mass renormalization, and together with the "bare" mass M_0 combines into the "observable" mass

$$M = M_0 + \frac{e^2}{2r_0}.$$

And the term $-e^2/(2r)$ in $m(r)$ leads to the shift

$$M \to M - \frac{e^2}{2r}$$

in the Schwarzschild metric (6.11), thus resulting in the Reissner – Nordström metric (6.35).

Not only do these considerations lead to the correct result, but they are essentially correct by themselves, differing in fact from the first, rigorous derivation in the following respect only: here we assume from the very beginning that $ad = 1$.

Interaction of Spin with Gravitational Field

In the present chapter, we use the term spin for brevity to mean the proper internal angular momentum of a classical particle, unrelated to its motion as a whole. In this sense one can talk for instance about the spin of a gyroscope installed on an Earth satellite (see section 7.2 below).

7.1 Spin-Orbit Interaction

We discuss here the interaction of spin **s** of a particle with its orbital angular momentum **l**, related to the motion of this particle in a centrally symmetric gravitational field. We assume that the field is weak, i.e. is described by the potential $\phi = -kM/r$, where, as usual, M is the mass of a source of a gravitational field. We are interested here in the interaction linear in spin **s**. Since the orbital angular momentum **l** of the particle is orthogonal both to its radius vector **r** and momentum **p**, the spin-orbit interaction, being a scalar, should be proportional to (**ls**). It is important that the scalar product (**ls**) of two axial vectors is a true scalar (but not a pseudoscalar), which is necessary in virtue of the invariance under the reflection of coordinates. Then, in a weak external field the spin-orbit interaction should be proportional to the magnitude of this field, i.e. to kM. After it, simple dimensional arguments dictate the form of the discussed interaction:

$$V_{ls} \sim \frac{kM}{mc^2 r^3} \, (\mathbf{ls}) \,, \tag{7.1}$$

where m is the mass of the particle.

We note the correspondence between (7.1) and the operator of spin-orbit interaction in a hydrogen-like ion with the charge of the nucleus Ze:

$$V_{ls}^{em} = \frac{Ze^2}{2m^2 c^2 r^3} \, (\mathbf{ls}) \tag{7.2}$$

(here the electron spin **s** and its orbital angular momentum **l** include the Planck constant $\hbar$ and have the dimension of action, as well as in our classical problem). Indeed, from the comparison of the Newton interaction kMm/r with the Coulomb one Ze^2/r (for the charges Ze and $-e$), the correspondence is obvious between kMm and Ze^2, and then between formulae (7.1) and (7.2). Moreover, the positive sign of the numerical constant in formula (7.2), originating in fact from the attracting Coulomb interaction, allows one to suppose that in the gravitational spin-orbit interaction (7.1), originating from the Newton attraction, a still unfound numerical factor will be positive as well. This is the case indeed.

Unfortunately, the explicit calculation of this factor is quite tedious.[1] Therefore we present here without derivation the complete formula for the gravitational spin-orbit interaction (*A.D. Fokker*, 1921):

$$V_{ls} = \frac{3}{2} \frac{kM}{mc^2 r^3} (\mathbf{ls}).$$

(7.3)

The equations of motion for spin are written via the Poisson brackets:

$$\frac{d\mathbf{s}}{dt} = \{V_{ls}, \mathbf{s}\}.$$

Using the Poisson brackets for the spin components $\{s_i, s_j\} = -\varepsilon_{ijk} s_k$ (they should have the same structure as those for the components of the orbital angular momentum), we obtain

$$\frac{d\mathbf{s}}{dt} = \frac{3}{2} \frac{kM}{mc^2 r^3} [\mathbf{l} \times \mathbf{s}].$$

Thus, the spin precesses with the angular velocity

$$\boldsymbol{\Omega} = \frac{3}{2} \frac{kM}{mc^2 r^3} \mathbf{l}.$$

(7.4)

Problem

7.1. In the gravitational field of a central body, a particle describes an ellipse with semi-major axis a and eccentricity e. Calculate the frequency of the spin precession averaged over the period. It is convenient to go from averaging over time to averaging over the angle ϕ by means of relations

$$\frac{dt}{T} = \frac{d\phi}{2\pi} \frac{(1 - e^2)^{3/2}}{(1 + e\cos\phi)^2}, \quad \frac{1}{r} = \frac{1 + e\cos\phi}{a(1 - e^2)}.$$

(7.5)

---

[1]Relatively economic calculation is described in the end of this chapter.

7.2 Spin-Spin Interaction

Now we discuss the interaction of the spin $\mathbf{s}$ of a probe particle with the spin $\mathbf{s}_0$ of a source of gravitational field. To linear approximation, $\mathbf{s}_0$ influences only the components h_{0n} of the gravitational field of a source. We note at once that $g_{0n} = \eta_{0n} + h_{0n} = h_{0n}$. One can easily demonstrate that $h_{0n} = h^{0n}$. Since we are interested here only in the effects due to the proper rotation of the source of the field, then all other components of $h_{\mu\nu}$ can be neglected.

At first we try to guess the general structure of the vector

$$\mathbf{g} = (h_{01}, h_{02}, h_{03}) = (g_{01}, g_{02}, g_{03}) \, .$$

It enters the interval ds^2 in the combination $g_{0n} dt dx^n$. Since the interval does not change sign under time reversal $dt \to -dt$ and is a true scalar (not a pseudoscalar), the vector $\mathbf{g}$ should change sign under time reversal together with dt, and should be a polar (not axial) vector together with dx^n. Due to the first requirement, $\mathbf{g}$ is proportional to $\mathbf{s}_0$; indeed, spin, like orbital angular momentum, changes sign under time reversal. However, spin is an axial vector, therefore, it should enter the expression for the polar vector $\mathbf{g}$ in the combination $\mathbf{r} \times \mathbf{s}_0$, where $\mathbf{r}$ is the radius-vector of the probe particle. Then, in the weak-field approximation $\mathbf{g}$ should be proportional to the Newton constant k. And finally, simple dimensional arguments prompt that

$$\mathbf{g} \sim k \, \frac{\mathbf{r} \times \mathbf{s}_0}{c^3 r^3} \, . \tag{7.6}$$

The direct calculation is not much more complicated. In the stationary case the equation for h_{0n} is (see (4.7)):

$$\Delta h_{0n} = 16 \, \pi \, k \, T_{0n} \, . \tag{7.7}$$

We assume that the internal motion in the source is nonrelativistic and rewrite the right-hand side of this equation as $16\pi k \rho v_n = -16\pi k \rho v^n$, where ρ is the mass density of the source, and v^n are the common, contravariant, components of the local velocity vector $\mathbf{v}$. It is clear now that equation (7.7) for the vector $\mathbf{g}$ coincides up to notations with the stationary equation for the vector-potential $\mathbf{A}$ in electrodynamics. Using the well-known solution of this last equation,[2] we find easily

$$\mathbf{g} = -2k \, \frac{\mathbf{r} \times \mathbf{s}_0}{c^3 r^3} \, . \tag{7.8}$$

Now we consider the motion of the vector of spin $\mathbf{s}$ of a probe particle in the gravitational field (7.8). We start with the covariant equation of motion for spin. The covariant vector of spin of a particle S_μ is defined in the flat space-time as follows. In the rest frame of the particle it has only space

---

[2]See, for instance, L.D. Landau and E.M. Lifshitz, *The Classical Theory of Fields*, § 44.

components, i.e. in this frame $S_\mu = (0, \mathbf{s})$, and in any other frame its components are found by means of the Lorentz transformation from the rest frame. The conservation of angular momentum in flat space-time means that the free covariant equation for spin is

$$\frac{dS^\mu}{d\tau} = 0 \tag{7.9}$$

(in the present chapter we denote the proper time by τ). Due to the principle of equivalence, in the gravitational field equation (7.9) goes over into

$$\frac{DS^\mu}{D\tau} = 0. \tag{7.10}$$

Now we rewrite equation (7.10) as

$$\frac{dS^\mu}{d\tau} + \Gamma^\mu_{\nu\tau} S^\nu u^\tau = 0, \tag{7.11}$$

and note that for the present problem of a nonrelativistic probe particle it is sufficient to put its velocity $\mathbf{v} = 0$. Then, with $S^\mu = (0, \mathbf{s})$ and $u^\mu = (1, 0)$, equation (7.11) simplifies to

$$\frac{ds^m}{dt} = -\Gamma^m_{n0} s^n = \Gamma_{m,n0} s^n = \frac{c}{2} (\nabla_n h_{0m} - \nabla_m h_{0n}) s^n,$$

or

$$\frac{d\mathbf{s}}{dt} = \frac{c}{2} \mathbf{s} \times [\boldsymbol{\nabla} \times \mathbf{g}].$$

Thus, spin $\mathbf{s}$ precesses in such a gravitational field with the angular velocity (*L. Schiff*, 1960)

$$\boldsymbol{\omega} = -\frac{c}{2} [\boldsymbol{\nabla} \times \mathbf{g}] = k \frac{3\mathbf{r}(\mathbf{r}\,\mathbf{s}_0) - r^2 \mathbf{s}_0}{c^2 r^5}. \tag{7.12}$$

The corresponding Hamiltonian of the spin-spin interaction appears as follows:

$$V_{ss} = (\boldsymbol{\omega}\mathbf{s}) = k \frac{3(\mathbf{r}\,\mathbf{s}_0)(\mathbf{r}\,\mathbf{s}) - r^2(\mathbf{s}_0\,\mathbf{s})}{c^2 r^5}. \tag{7.13}$$

Let us note that the spin precesses in such a way as if it were considered in a reference frame rotating with the angular velocity $-\boldsymbol{\omega}$ with respect to the inertial frame where spin is at rest. In this sense one can talk about "dragging" the inertial frame with the angular velocity $-\boldsymbol{\omega}$ caused by the proper angular momentum of the source of gravitational field.

Problems

7.2. Prove relation $h_{0n} = h^{0n}$.

7.3. A thin spherical shell of radius R rotates with an angular velocity Ω. Its total mass is distributed uniformly. Find the metric outside and inside the shell, assuming that the deviation of the metric from the flat one is small. Find the angular velocity ω of dragging inertial frames inside the shell.

7.4. Find the contribution to the deviation of beam of light due to the rotation of gravitating center. Assume that the plane of motion of the beam is orthogonal to the axis of rotation of the center.

7.5. A satellite with gyroscope is on an orbit around the Earth. Estimate the frequency of the gyroscope precession 1) due to the spin-orbit interaction, 2) due to the spin-spin interaction with the proper angular momentum $\mathbf{s}_0$ of the Earth rotation. How should one orient the gyroscope axis with respect to the plane of the satellite orbit, and the plane of orbit with respect to $\mathbf{s}_0$, to amplify in a maximum way the relative contribution of the second effect with respect to the first one?

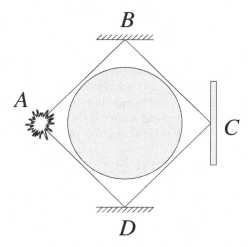

Fig. 7.1. Shift of interference fringes

7.6. Beams of light emitted by a source situated at point A, propagate along the paths ABC and ADC, and interfere on the screen situated at point C (see Fig. 7.1). At the center of the square $ABCD$ there is a rotating body with the rotation axis orthogonal to the plane of the square. Estimate numerically the shift of the interference fringes due to the rotation, if the rotating body is the Earth, and the side of the square equals the Earth diameter (*I.B. Khriplovich, O.L. Zhizhimov*, 1980).

7.3 Orbit Precession Due to Rotation of Central Body

Rotation of a central body causes the precession not only of the spin of a particle, but the orbit of this particle as well. Not only the perihelion, i.e. the Runge – Lenz vector

$$\mathbf{A} = \frac{1}{m} [\mathbf{p} \times \mathbf{l}] - \frac{kmM\mathbf{r}}{r}, \tag{7.14}$$

precesses now, as this is the case in the central field due to the nonlinear correction to the potential. In the present case, due to the noncentral correction to the field, the orbital angular momentum l is not conserved also. It precesses, and with it the plane of the orbit precesses as well since the normal to it is directed along l (*J. Lense, H. Thirring*, 1918).

The correction to the Lagrangian $L = -mds/dt$ of a particle with mass m, due to nonvanishing vector **g**, is

$$\delta L = -mc(\mathbf{gv}) = -\frac{2km}{c^2 r^3} ([\mathbf{r} \times \mathbf{v}] \, \mathbf{s}_0).$$

The corresponding correction to the particle Hamiltonian is

$$V_{ls1} = \frac{2k}{c^2 r^3} (\mathbf{s}_0 \, \mathbf{l}). \tag{7.15}$$

Let us draw attention to the analogy between the gravitational effects, discussed in the present chapter, and effects from atomic physics. For the spin-orbit interaction (7.3) this analogy has been already mentioned. In the present case, the spin-spin interaction (7.13) is an obvious analogue of the hyperfine spin-spin interaction,[3] and (7.15) corresponds to the hyperfine interaction of the electron orbital angular momentum with the nuclear spin.

Equation of motion for the orbital angular momentum of the particle appears as follows:

$$\frac{d\mathbf{l}}{dt} = \{V_{ls1}, \mathbf{l}\} = \frac{2k}{c^2 r^3} [\mathbf{s}_0 \times \mathbf{l}],$$

i.e. the orbital angular momentum of the particle, together with the plane of its orbit, precesses with the angular velocity

$$\boldsymbol{\omega}_1 = \frac{2k}{c^2 r^3} \mathbf{s}_0.$$

The angular velocity averaged over the period is

$$\langle \boldsymbol{\omega}_1 \rangle = \frac{2k}{c^2 a^3 (1 - e^2)^{3/2}} \mathbf{s}_0. \tag{7.16}$$

The time derivative of the Runge – Lenz vector (7.14) is calculated in an analogous way. Its averaged angular velocity is

---

[3]Of course, we do not mean here atomic s-states where the last interaction is of the contact nature, i.e. is proportional to $\delta(\mathbf{r})$ instead of $1/r^3$.

$$\langle \boldsymbol{\omega}_2 \rangle = \frac{2k}{c^2 a^3 (1 - e^2)^{3/2}} \left[\mathbf{s}_0 - 3 \frac{\mathbf{l}(\mathbf{l}\mathbf{s}_0)}{l^2} \right] . \qquad (7.17)$$

Obviously, it can be said that the plane of the orbit, together with l, also precesses with the averaged angular velocity $\langle \boldsymbol{\omega}_2 \rangle$. In other words, $\langle \boldsymbol{\omega}_2 \rangle$ is the angular velocity of the precession in space of the ellipse of the orbit as a whole.

Problems

7.7. Prove formulae (7.16) and (7.17).

7.8. What is the form of the gravitational spin-orbit and spin-spin interaction in the two-body problem for particles with different masses m_1, m_2 and spins $\mathbf{s}_1$, $\mathbf{s}_2$?

7.4 Equations of Motion of Spin in Electromagnetic Field

In the next section the general problem of the spin precession in an external gravitational field will be reduced to the analogous problem for the case of an external electromagnetic field. The equations of motion for spin of a relativistic particle in electromagnetic field are not directly related to GR, and besides, they are well known.[4] However, at least to make the presentation coherent, we will consider in this section just the problem referring to the electromagnetic field.

We start with the spin precession for a nonrelativistic charged particle. The equation that describes this precession is well known:

$$\dot{\mathbf{s}} = \frac{eg}{2m} \left[\mathbf{s} \times \mathbf{B} \right] . \qquad (7.18)$$

Here $\mathbf{B}$ is an external magnetic field, e and m are the charge and mass of the particle, g is its gyromagnetic ratio (for electron $g \approx 2$). In other words, the spin precesses around the direction of magnetic field with the frequency $-(eg/2m)\mathbf{B}$. In the same nonrelativistic limit the velocity precesses around the direction of $\mathbf{B}$ with the frequency $-(e/m)\mathbf{B}$:

$$\dot{\mathbf{v}} = \frac{e}{m} \left[\mathbf{v} \times \mathbf{B} \right] .$$

Thus, for $g = 2$ spin and velocity precess with the same frequency, so that the angle between them is conserved.

Now we are going over to the relativistic generalization of equation (7.18). We will use here at first the four-dimensional vector of spin S_μ, already discussed in section 7.2. In the reference frame where the particle moves with

---

[4]See, for instance, V.B. Berestetsky, E.M. Lifshitz, and L.P. Pitaevsky, *Quantum Electrodynamics*, § 41.

velocity $\mathbf{v}$, the vector S_μ is constructed from $(0, \mathbf{s})$ by means of the Lorentz transformation, so that here

$$S_0 = \gamma \mathbf{vs}, \quad \mathbf{S} = \mathbf{s} + \frac{\gamma^2 \mathbf{v(vs)}}{\gamma + 1}. \qquad (7.19)$$

Then, just by definition of S_μ, the following identities take place:

$$S_\mu S_\mu = -\mathbf{s}^2 \, (= \text{const}), \quad S_\mu u_\mu = 0; \qquad (7.20)$$

as usual, here u_μ is the four-velocity.

The right-hand side of the equation for $dS_\mu/d\tau$ should be linear and homogeneous both in the electromagnetic field strength $F_{\mu\nu}$, and in the same four-vector S_μ, and may depend also on u_μ. In virtue of the first identity (7.20), the right-hand side should be four-dimensionally orthogonal to S_μ. Therefore, the general structure of the equation we are looking for, is

$$\frac{dS_\mu}{d\tau} = \alpha F_{\mu\nu} S_\nu + \beta u_\mu F_{\nu\lambda} u_\nu S_\lambda. \qquad (7.21)$$

Comparing the nonrelativistic limit of this equation with (7.18), we find

$$\alpha = \frac{eg}{2m}.$$

Now we take into account the second identity (7.20), which after differentiation in τ gives

$$u_\mu \frac{dS_\mu}{d\tau} = -S_\mu \frac{du_\mu}{d\tau},$$

and recall the classical equation of motion for a charge:

$$m \frac{du_\mu}{d\tau} = e F_{\mu\nu} u_\nu. \qquad (7.22)$$

Then, multiplying equation (7.21) by u_μ, we obtain

$$\beta = -\frac{e}{2m} (g - 2).$$

Thus, the covariant equation of motion for spin is

$$\frac{dS_\mu}{d\tau} = \frac{eg}{2m} F_{\mu\nu} S_\nu - \frac{e}{2m} (g - 2) u_\mu F_{\nu\lambda} u_\nu S_\lambda \qquad (7.23)$$

(*Ya.I. Frenkel*, 1926; *V. Bargman, L. Michel, V. Telegdi*, 1959).

Let us discuss the limits of applicability for this equation.

Of course, typical distances at which the trajectory changes (for instance, the Larmor radius in a magnetic field) should be large as compared to the de Broglie wave length $\hbar/p$ of the elementary particle. Then, the external field itself should not change essentially at the distances on the order of both the de Broglie wave length $\hbar/p$ and the Compton wave length $\hbar/(mc)$ of the particle. In particular, if the last condition does not hold, the scatter of velocities in the rest frame is not small as compared to c, and one cannot use in this frame the nonrelativistic formulae.

Besides, if the external field changes rapidly, the motion of spin will be influenced by interaction of higher electromagnetic multipoles of the particle with field gradients. For a particle of spin 1/2 higher multipoles are absent, and the gradient-dependent effects are due to finite form factors of the particle. These effects start here at least in second order in field gradients and usually are negligible.

At last, in equation (7.23) we confine to effects of first order in the external field. This approximation relies in fact on the implicit assumption that the first-order interaction with the external field is less than the excitation energy of the spinning system. Usually this assumption is true and the first-order equation (7.23) is valid. Still, one can easily point out situations when this is not the case. To be definite, let us consider the hydrogen-like ion $^3\mathrm{He}^+$ in the ground s-state with the total spin $F = 1$. It can be easily demonstrated that an already quite moderate external magnetic field is sufficient to break the hyperfine interaction between the electron and nuclear magnetic moments (a sort of Paschen – Back effect). Then, instead of a precession of the total spin $\mathbf{F}$ of the ion, which should be described by equations (7.18) or (7.23) with a corresponding ion g-factor, we will have a separate precession of the decoupled electron and nuclear spins.

Let us go back now to equation (7.23). We note that for $g = 2$ and in the absence of electric field, its zeroth component reduces to

$$\frac{dS_0}{d\tau} = 0.$$

Taking into account definition (7.19) for S_0 and the fact that in a magnetic field a particle energy remains constant, we find immediately that the projection of spin $\mathbf{s}$ onto velocity, so-called helicity, is conserved.

We will obtain now the relativistic equation for the three-dimensional vector of spin $\mathbf{s}$, that directly describes the internal angular momentum of a particle in its "momentary" rest frame. This equation can be derived from (7.23) using relations (7.19), together with the equations of motion for a charge in external field. It will require, however, quite tedious calculations. Therefore, we choose another way, somewhat more simple and much more instructive.

First, we transform equation (7.18) from the comoving inertial frame, where the particle is at rest, into the laboratory one. The magnetic field $\mathbf{B}'$ in the rest frame is expressed via the electric and magnetic fields $\mathbf{E}$ and $\mathbf{B}$ given in the laboratory frame, as follows:

$$\mathbf{B}' = \gamma\mathbf{B} - \frac{\gamma^2}{\gamma + 1}\,\mathbf{v}(\mathbf{vB}) - \gamma\mathbf{v} \times \mathbf{E}.$$

This expression can be easily checked by comparing it component by component with the transformation of magnetic field for two cases: when this field is parallel to the velocity and orthogonal to it, respectively. Then one should take into account that the frequency in the laboratory time t is γ times smaller than the frequency in the laboratory time τ (indeed, $d/dt = d\tau/dt \cdot d/d\tau = \gamma^{-1} d/d\tau$). Found in this way contribution to the precession frequency is

$$\boldsymbol{\omega}_g = -\frac{eg}{2m} \left[\mathbf{B} - \frac{\gamma}{\gamma+1} \mathbf{v}(\mathbf{vB}) - \mathbf{v} \times \mathbf{E} \right].$$

However it is clear from equation (7.23) that spin precesses even if $g = 0$. To elucidate the origin of this effect, the so-called Thomas precession (*L. Thomas*, 1926), we consider two successive Lorentz transformations: at first from the laboratory frame S into the frame S' that moves with the velocity $\mathbf{v}$ with respect to S, and then from S' into the frame S'' that moves with respect to S' with the infinitesimal velocity $d\mathbf{v}$. Let us recall in this connection the following fact related to usual three-dimensional rotations: the result of two successive rotations with respect to noncollinear axes $\mathbf{n}_1$ and $\mathbf{n}_2$ contains in particular a rotation around the axis directed along their vector product $\mathbf{n}_1 \times \mathbf{n}_2$. Now it is only natural to assume that the result of the above successive Lorentz transformations will contain in particular a usual rotation around the axis directed along $d\mathbf{v} \times \mathbf{v}$. In result, spin in the rest frame will rotate in the opposite direction by an angle which we denote by $\varkappa [d\mathbf{v} \times \mathbf{v}]$. Here $\varkappa$ is some numerical factor to be determined below. It depends generally speaking on the particle energy.

This is in fact the Thomas precession. Its frequency in the proper time τ is

$$\boldsymbol{\omega}'_T = \varkappa [d\mathbf{v}/d\tau \times \mathbf{v}] = \kappa \frac{e}{m} [\mathbf{E}' \times \mathbf{v}].$$

Now we transform the electric field $\mathbf{E}'$ from the proper frame into the laboratory one, as it was done above for the magnetic field $\mathbf{B}'$, and go over also from the proper time τ to t. In result, the frequency of the Thomas precession in the laboratory frame is

$$\boldsymbol{\omega}_T = \varkappa \frac{e}{m} \left[\left(\mathbf{E} - \frac{\gamma}{\gamma+1} \mathbf{v}(\mathbf{vE}) + \mathbf{v} \times \mathbf{B} \right) \times \mathbf{v} \right]$$

$$= -\varkappa \frac{e}{m} \left[\mathbf{v} \times \mathbf{E} - v^2 \mathbf{B} + \mathbf{v}(\mathbf{vB}) \right].$$

To find the coefficient $\varkappa$, we recall that in a magnetic field, for $g = 2$ the projection of spin onto the velocity is conserved. In other words, in this case the total frequency of the spin precession $\boldsymbol{\omega} = \boldsymbol{\omega}_g + \boldsymbol{\omega}_T$ coincides with the frequency of the velocity precession which is well known to be

$$\boldsymbol{\omega}_v = -\frac{e}{m\gamma} \mathbf{B}.$$

From this we find easily that $\varkappa = \gamma/(\gamma + 1)$. Correspondingly, the relativistic equation of motion for the three-dimensional vector of spin $\mathbf{s}$ in external electromagnetic field is

$$\frac{d\mathbf{s}}{dt} = (\boldsymbol{\omega}_g + \boldsymbol{\omega}_T) \times \mathbf{s} = \frac{e}{2m} \left\{ \left(g - 2 + \frac{2}{\gamma} \right) [\mathbf{s} \times \mathbf{B}] \right.$$

$$\left. - (g - 2) \frac{\gamma}{\gamma + 1} [\mathbf{s} \times \mathbf{v}](\mathbf{v}\mathbf{B}) - \left(g - \frac{2\gamma}{\gamma + 1} \right) [\mathbf{s} \times [\mathbf{v} \times \mathbf{E}]] \right\}. \qquad (7.24)$$

Problems

7.9. Derive equation (7.24) directly from (7.23).

7.10. Obtain the Hamiltonian of spin-orbit interaction in hydrogen atom from equation (7.24).

7.11. Derive equation of motion of the quadrupole moment of a relativistic particle in homogeneous electric and magnetic fields. In the rest frame, the operator of quadrupole moment is

$$q_{mn} = \frac{3q}{2s(2s-1)} \left[s_m s_n + s_n s_m - \frac{2}{3} s(s+1)\delta_{mn} \right].$$

7.5 Equations of Motion of Spin in Gravitational Field

It has been pointed out in section 7.2 that the covariant equation of motion for spin is

$$\frac{DS^\mu}{D\tau} = 0. \qquad (7.25)$$

However, the notion of spin is directly related to the group of rotations. It is only natural, therefore, to describe spin in the local Lorentz coordinate frame using the tetrad formalism (see section 3.1). The tetrad components of spin

$$S^a = S^\mu e^a_\mu$$

(by the first letters of the Latin alphabet, a, b, c, d, we label here and below four-dimensional tetrad indices) behave as vectors under Lorentz transformations of the locally inertial frame. However, they do not change under generally covariant transformations $x^\mu = f^\mu(x')$. In other words, the four components S^a are world scalars. Therefore, in virtue of relation (7.10), the equations for them appear as follows:

$$\frac{DS^a}{D\tau} = \frac{dS^a}{d\tau} = S^\mu e^a_{\mu;\,\nu} u^\nu = \eta^{ab} \gamma_{bcd} u^d S^c. \qquad (7.26)$$

The covariant derivative of a tetrad is by definition

$$e^a_{\mu;\nu} = \partial_\nu e^a_\mu - \Gamma^\varkappa_{\mu\nu} e^a_\varkappa ,$$

and the quantity

$$\gamma_{abc} = e_{a\mu;\nu} e^\mu_b e^\nu_c \qquad (7.27)$$

is called the Ricci rotation coefficient. By means of covariant differentiation of the identity $e_{a\mu} e^\mu_b = \eta_{ab}$, one can easily demonstrate that these coefficients are antisymmetric in the first pair of indices:

$$\gamma_{abc} = -\gamma_{bac} . \qquad (7.28)$$

Of course, the equations for the tetrad components of a 4-velocity look exactly in the same way as those for spin:

$$\frac{du^a}{d\tau} = \eta^{ab} \gamma_{bcd}\, u^d u^c . \qquad (7.29)$$

The meaning of equations (7.26) and (7.29) is clear: the tetrad components of both vectors vary in the same way since their variation is due only to the rotation of the local Lorentz frame.

There is a remarkable similarity between the discussed problem and the special case of $g = 2$ in electrodynamics. According to equations (7.23) and (7.22), the four-dimensional spin and four-dimensional velocity of a charged particle with the gyromagnetic ratio $g = 2$ precess with the same angular velocity:

$$\frac{dS_a}{d\tau} = \frac{e}{m} F_{ab} S^b, \quad \frac{du_a}{d\tau} = \frac{e}{m} F_{ab} u^b.$$

In other words, the obvious correspondence takes place:

$$\frac{e}{m} F_{ab} \longleftrightarrow \gamma_{abc} u^c. \qquad (7.30)$$

It allows us to derive the precession frequency $\boldsymbol{\omega}$ of a three-dimensional vector of spin $\mathbf{s}$ in an external gravitational field from expression (7.24) by means of the simple substitution

$$\frac{e}{m} B_i \longrightarrow -\frac{1}{2} \epsilon_{ikl} \gamma_{klc} u^c; \quad \frac{e}{m} E_i \longrightarrow \gamma_{0ic} u^c. \qquad (7.31)$$

Thus, this frequency is (*I.B. Khriplovich, A.A. Pomeransky, 1998*)

$$\omega_i = \epsilon_{ikl} \left(\frac{1}{2} \gamma_{klc} + \frac{u^k}{u^0 + 1} \gamma_{0lc} \right) \frac{u^c}{u^0_w} . \qquad (7.32)$$

The factor $1/u^0_w$ in expression (7.32) is due to the transition in the left-hand side of equation (7.26) to differentiating over the world time t:

$$\frac{d}{d\tau} = \frac{dt}{d\tau}\frac{d}{dt} = u_w^0\frac{d}{dt}.$$

We supply here u_w^0 with the subscript w to indicate that this is the world, but not the tetrad, component of 4-velocity. All other indices in (7.32) are tetrad ones, $c = 0, 1, 2, 3$, $i, k, l = 1, 2, 3$.

However, in some respect the first-order spin interaction with a gravitational field differs essentially from that with an electromagnetic field. In the case of an electromagnetic field, the interaction depends, generally speaking, on a free phenomenological parameter, g-factor. Moreover, if one allows for the violation of invariance both under the reflection of space coordinates and under time reversal, one more parameter arises in the case of electromagnetic interaction, the value of the electric dipole moment of the particle. The point is that both magnetic and electric dipole moments interact with the electromagnetic field strength, so that this interaction is gauge-invariant for any value of these moments. Only the spin-independent interaction with the electromagnetic vector potential is fixed by the charge conservation and gauge invariance. On the contrary, the Ricci rotation coefficients γ_{abc} entering the gravitational first-order spin interaction (7.26), as distinct from the Riemann tensor, are noncovariant. Therefore, the discussed interaction of spin with gravitational field is fixed in unique way by the law of angular momentum conservation in flat space-time taken together with the equivalence principle, and thus it contains no free parameters (*L.D. Landau*). On the other hand, it is no surprise that the precession frequency $\boldsymbol{\omega}$ depends not on the Riemann tensor, but on the rotation coefficients. Of course, this frequency should not be a tensor: it is sufficient to recall that a spin, which is at rest in an inertial reference frame, precesses in a rotating one.

One can check easily that in the weak-field approximation where

$$g_{\mu\nu} = \eta_{\mu\nu} + h_{\mu\nu}, \quad |h_{\mu\nu}| \ll 1,$$

there is no difference between the tetrad and world indices in $e_{a\mu}$, and the tetrad appears as follows:

$$e_{\mu\nu} = \eta_{\mu\nu} + \tilde{e}_{\mu\nu}, \quad |\tilde{e}_{\mu\nu}| \ll 1.$$

Relation between the tetrads and metric

$$e_{a\mu}e_{b\nu}\eta^{ab} = g_{\mu\nu}$$

in the weak-field approximation reduces to

$$\tilde{e}_{\mu\nu} + \tilde{e}_{\nu\mu} = h_{\mu\nu}.$$

Under the demand that tetrads are expressed via metric only, one arrives at the so-called symmetric gauge for the tetrads where

$$\tilde{e}_{\mu\nu} = \frac{1}{2}h_{\mu\nu}.$$

Then in the weak-field approximation the Ricci coefficients are:

$$\gamma_{abc} = \frac{1}{2} \left(h_{bc,a} - h_{ac,b} \right).$$ (7.33)

Now, with relations (7.32) and (7.33) one can solve, for instance, in an elementary way the problems of spin-orbit and spin-spin interactions for arbitrary particle velocities. The combination of a high velocity for a spinning particle with a weak gravitational field refers obviously to a scattering problem. Another possible application is to a spinning particle bound by other forces, for instance, by electromagnetic ones, when we are looking for the correction to the precession frequency due to the gravitational interaction. So, let us consider the spin-orbit and spin-spin problems.

We start with the spin-orbit interaction. In the centrally symmetric field created by a mass M, the metric is

$$h_{00} = -\frac{r_g}{r} = -\frac{2kM}{r}, \quad h_{mn} = -\frac{r_g}{r} \delta_{mn} = -\frac{2kM}{r} \delta_{mn}.$$ (7.34)

Here the nonvanishing Ricci coefficients are

$$\gamma_{ijk} = \frac{kM}{r^3} \left(\delta_{jk} r_i - \delta_{ik} r_j \right), \quad \gamma_{0i0} = -\frac{kM}{r^3} r_i.$$ (7.35)

Plugging these expressions into formula (7.32) yields the following result for the precession frequency:

$$\boldsymbol{\omega}_{ls} = \frac{2\gamma + 1}{\gamma + 1} \frac{kM}{r^3} \mathbf{v} \times \mathbf{r}.$$ (7.36)

In the limit of low velocities, $\gamma \to 1$, the answer goes over into the classical result (7.4).

And now the spin-orbit interaction. Using expression (7.8) for the components of the metric due to the spin $\mathbf{s}_0$ of central body, we find the nonvanishing Ricci coefficients:

$$\gamma_{ij0} = k \left(\nabla_i \frac{[\mathbf{s}_0 \times \mathbf{r}]_j}{r^3} - \nabla_j \frac{[\mathbf{s}_0 \times \mathbf{r}]_i}{r^3} \right), \quad \gamma_{0ij} = -k \nabla_i \frac{[\mathbf{s}_0 \times \mathbf{r}]_j}{r^3}.$$ (7.37)

The frequency of the spin-spin precession is

$$\boldsymbol{\omega}_{ss} = k \left(2 - \frac{1}{\gamma} \right) (\mathbf{s}_0 \boldsymbol{\nabla}) \boldsymbol{\nabla} \frac{1}{r}$$

$$- k \frac{\gamma}{\gamma + 1} \left[\mathbf{v}(\mathbf{s}_0 \boldsymbol{\nabla}) - \mathbf{s}_0(\mathbf{v}\boldsymbol{\nabla}) + (\mathbf{v}\mathbf{s}_0)\boldsymbol{\nabla} \right] (\mathbf{v}\boldsymbol{\nabla}) \frac{1}{r}.$$ (7.38)

In the low-velocity limit this formula also goes over into the classical result (7.12).

Problems

7.12. Prove identity

$$e_{a\mu;\,\lambda} - \gamma_{ab\lambda}e_{\mu}^{b} = \partial_{\lambda}e_{a\mu} - \Gamma^{\rho}_{\mu\lambda}e_{a\rho} - \gamma_{ab\lambda}e_{\mu}^{b} = 0$$

for the tetrad. Compare it with identity

$$g_{\mu\nu;\,\lambda} = \partial_{\lambda}g_{\mu\nu} - \Gamma^{\rho}_{\mu\lambda}g_{\rho\nu} - \Gamma^{\rho}_{\lambda\nu}g_{\mu\rho} = 0$$

for the metric tensor.

7.13. Is $e^{a}_{\mu;\,\nu}$ a tensor in the Riemann space?

7.14. Find the frequency of spin precession in the Schwarzschild field for circular orbits (beyond the weak-field approximation) (*T.A. Apostolatos*, 1996).

8

Gravitational Waves

8.1 Free Gravitational Wave

In this chapter (as well as in the previous one) we do not go beyond the linear approximation to the Einstein equations.[1] In the linear approximation, under the auxiliary harmonic condition (4.6), the gravitational field is described by equation (4.7). The solutions of the corresponding free equation are gravitational waves.

We demonstrate first of all that for an arbitrary weak field $h_{\mu\nu}(x)$ one can always choose such a coordinate transformation

$$x'^\mu = x^\mu + \varepsilon^\mu(x),$$

after which the transformed field $h'_{\mu\nu}(x')$ will satisfy condition (4.6). Indeed, in virtue of the transformation law (3.6) as applied to the metric tensor $g_{\mu\nu} = \eta_{\mu\nu} + h_{\mu\nu}$, the following relation takes place

$$\eta_{\mu\nu} + h'_{\mu\nu}(x') = \frac{\partial x^\rho}{\partial x'^\mu}\frac{\partial x^\tau}{\partial x'^\nu}[\eta_{\rho\tau} + h_{\rho\tau}(x)] = \eta_{\mu\nu} + h_{\mu\nu}(x) - \partial_\mu\varepsilon_\nu(x) - \partial_\nu\varepsilon_\mu(x),$$

or

$$h'_{\mu\nu}(x) = h_{\mu\nu}(x) - \partial_\mu\varepsilon_\nu(x) - \partial_\nu\varepsilon_\mu(x). \tag{8.1}$$

Let us note that since both $h_{\mu\nu}$ and ε_μ are small, there is no reason to distinguish in the arguments of these functions x and x'. Now

$$\partial_\mu h'_{\mu\nu}(x) - \frac{1}{2}\partial_\nu h'_{\mu\mu}(x) = \partial_\mu h_{\mu\nu}(x) - \frac{1}{2}\partial_\nu h_{\mu\mu}(x) - \Box\varepsilon_\nu(x).$$

Thus, for an arbitrary initial $h_{\mu\nu}(x)$, by choosing the vector parameters $\varepsilon_\nu(x)$ in such a way that they satisfy the equation

---

[1]Only in section (8.6) we discuss a weak gravitational wave radiated during the motion in a strong external gravitational field.

$$\Box \varepsilon_\nu(x) = \partial_\mu h_{\mu\nu}(x) - \frac{1}{2} \partial_\nu h_{\mu\mu}(x)\,, \qquad (8.2)$$

one can always make $h'_{\mu\nu}(x)$ satisfying the harmonic condition

$$\partial_\mu h'_{\mu\nu}(x) - \frac{1}{2} \partial_\nu h'_{\mu\mu}(x) = 0\,.$$

However, this condition still does not fix the reference frame uniquely. Obviously, one can perform over the field $h_{\mu\nu}(x)$, for which the harmonic condition is valid, a new coordinate transformation (8.1) with parameters $\varepsilon_\nu(x)$ satisfying the condition

$$\Box \varepsilon_\nu(x) = 0\,.$$

The harmonic condition, combined with the possibility of this last additional coordinate transformation, allows one to fix the tensor structure of a plane wave. So, let $h_{\mu\nu}(x) = e_{\mu\nu} e^{-ikx}$. In virtue of wave equation (4.7), the 4-vector k_μ satisfies the condition $k^2 = 0$. The harmonic condition for the polarization tensor $e_{\mu\nu}$ appears as follows:

$$k_\mu e_{\mu\nu} - \frac{1}{2} k_\nu e_{\mu\mu} = 0\,. \qquad (8.3)$$

We choose the wave vector as $k_\mu = \omega(1,0,0,1)$. Then the components $\nu = a = 1,2$ of equation (8.3) give

$$e_{a0} = e_{a3}\,.$$

The sum of the components with $\nu = 0$ and $\nu = 3$ results in

$$e_{aa} = e_{11} + e_{22} = 0\,.$$

Then it follows from the component with $\nu = 0$ that

$$e_{03} = \frac{1}{2}\left(e_{00} + e_{33}\right)\,.$$

Now we perform the additional transformation with parameters $\varepsilon_\nu(x) = i\varepsilon_\nu e^{-ikx}$:

$$e'_{\mu\nu} = e_{\mu\nu} - k_\mu \varepsilon_\nu - k_\nu \varepsilon_\mu\,.$$

Putting in it

$$\varepsilon_a = \frac{e_{a3}}{\omega} \ (a = 1,2)\,, \quad \varepsilon_0 = \frac{e_{00}}{2\omega}\,, \quad \varepsilon_3 = \frac{e_{33}}{2\omega}\,,$$

we turn to zero

$$e'_{a3}\,, \quad e'_{00}\,, \quad e'_{33}\,,$$

respectively. In the result, the polarization tensor has only two independent components:

$$e_{11} = -e_{22}, \quad e_{12} = e_{21}$$

(we omit the primes now).

Let us consider how the 2×2 matrix

$$\begin{pmatrix} e_{11} & e_{12} \\ e_{12} & -e_{11} \end{pmatrix}$$

transforms under the rotation by angle ϕ around the z axis. This transformation $e'_{ab} = O_{ac}O_{bd}e_{cd}$, where

$$O = \begin{pmatrix} \cos\phi & \sin\phi \\ -\sin\phi & \cos\phi \end{pmatrix},$$

is conveniently rewritten as $e' = O\,e\,O^T$. After it we find easily

$$e'_{11} = \cos 2\phi\, e_{11} + \sin 2\phi\, e_{12}, \quad e'_{12} = -\sin 2\phi\, e_{11} + \cos 2\phi\, e_{12}.$$

Now we go over from the linear polarizations e_{11}, e_{12} to the circular ones

$$e_{\pm} = \frac{-e_{11} \mp i e_{12}}{\sqrt{2}}.$$

For $e_{\pm}$ this transformation appears as follows:

$$e'_{\pm} = e^{\mp 2i\phi}\, e_{\pm}. \tag{8.4}$$

By the analogy with the quantum of electromagnetic field, the photon, one introduces the notion of graviton, the quantum of the gravitational field. The transformation law (8.4) means that the projection of the total angular momentum of a graviton onto the direction of its momentum, i.e. onto z axis, equals ± 2. And since the projection of the orbital angular momentum onto the momentum vanishes identically, it means that the projection of the graviton spin onto the direction of its motion, i.e. its helicity, is ± 2. Let us recall that the photon helicity is ± 1. The more general assertion is valid: for any (nonzero) spin s of a massless particle, this particle has only two polarization states, with helicities $\pm s$.

One more remark concerning the graviton. In chapter 4 it was demonstrated that the requirement of general covariance fixes strictly the form of the second-order equation for the gravitational field. In this way, as strictly fixed is the linear approximation to this equation, and it follows explicitly from the linear approximation that the gravitational field is massless. The only additional assumption made, that of the absence of the cosmological constant, is in fact inessential for this conclusion. It can be easily seen that in the weak-field limit the cosmological term in the wave equation reduces to a constant, but not to the additive term $-m^2 h_{\mu\nu}$, which would correspond to a finite mass. Thus, in the generally covariant second-order equation there is no place for the nonvanishing graviton mass. There are no experimental indications of the finite mass of the gravitational field. Its detection would mean a cardinal going beyond the framework of GR.

Problems

8.1. Find the components of the Riemann tensor for a plane gravitational wave propagating along the z axis.

8.2. Find how the relative distance between two particles changes with time under the action of a plane gravitational wave propagating along the z axis. Assume that the particles were initially in the plane xy.

8.3. Find the frequency of spin precession in the field of a plane gravitational wave propagating along the z axis.

8.2 Radiation of Gravitational Waves

We come back now to equation (4.7). Taking its trace and thus expressing $T_{\lambda\lambda}$ through $h_{\lambda\lambda}$, we rewrite this equation as

$$-\Box\psi_{\mu\nu} = 16\pi k\, T_{\mu\nu}\,, \qquad (8.5)$$

where $\psi_{\mu\nu} = h_{\mu\nu} - \frac{1}{2}\eta_{\mu\nu}h_{\lambda\lambda}$. It is clear from equation (8.5) that the harmonic condition $\partial_\mu\psi_{\mu\nu} = 0$ agrees in a natural way with the conservation law $\partial_\mu T_{\mu\nu} = 0$. It should be noted that though equation (8.5) is linear in the field of a gravitational wave, it is not necessarily linear in a gravitational field in general. Indeed, it is sufficient to consider a simple case when the source is a system of nonrelativistic particles bound by gravitational forces. Obviously, without the account for the stress tensor quadratic in the gravitational potential ϕ, $T_{\mu\nu}$ will not be conserved.

The retarded solution of equation (8.5) is

$$\psi_{\mu\nu}(\mathbf{R},t) = -4k \int \frac{d\mathbf{r}'\, T_{\mu\nu}(\mathbf{r}',t - |\mathbf{R}-\mathbf{r}'|)}{|\mathbf{R}-\mathbf{r}'|}\,.$$

As it was shown in section 8.1, in the wave zone $h_{\lambda\lambda} \to 0$, so that in this region $\psi_{\mu\nu} \to h_{\mu\nu}$ and, moreover, $h_{\mu\nu}$ has pure space components only. Besides, here one can change in the denominator for $R \gg r'$, as usual, $|\mathbf{R}-\mathbf{r}'| \to R$. Thus, at large distances we have

$$h_{mn}(\mathbf{R},t) = -\frac{4k}{R} \int d\mathbf{r}'\, T_{mn}(\mathbf{r}',t - |\mathbf{R}-\mathbf{r}'|)\,.$$

For a system of nonrelativistic particles, the integrand in the right-hand side is conveniently transformed in such a way that instead of T_{mn}, which depends on details of the motion and interaction of these particles, only their mass distribution enters this integrand. To this end, we at first integrate with the weight x_k the conservation law $\partial_\mu T_{\mu n} = 0$:

$$\int d\mathbf{r} \, x_k (\partial_0 T_{0n} + \partial_m T_{mn}) = \partial_t \int d\mathbf{r} \, x_k \, T_{0n} - \int d\mathbf{r} \, T_{kn} = 0 \, .$$

With the account for the symmetry $T_{kn} = T_{nk}$, we rewrite the obtained relation as

$$\int d\mathbf{r} \, T_{kn} = \frac{1}{2} \partial_t \int d\mathbf{r} \, (x_k \, T_{0n} + x_n \, T_{0k}) \, .$$

In the analogous way, integrating with the weight $x_k x_l$ the conservation law $\partial_\mu T_{\mu 0} = 0$, we obtain

$$\int d\mathbf{r} \, x_k \, x_l \, (\partial_0 T_{00} + \partial_m T_{m0}) = \partial_t \int d\mathbf{r} \, x_k \, x_l \, T_{00} - \int d\mathbf{r} \, (x_k \, T_{0l} + x_l \, T_{0k}) = 0 \, .$$

Thus,

$$h_{mn} = -\frac{2k}{R} \partial_t^2 \int d\mathbf{r} \, x_m \, x_n \, T_{00} \, .$$

We recall now that for a nonrelativistic system T_{00} coincides with the mass density ρ, and that $h_{nn} = 0$. Then the result is expressed through the quadrupole moment q_{mn} of the mass distribution:

$$h_{mn} = -\frac{2k}{3R} \ddot{q}_{mn} \, , \quad q_{mn} = \int d\mathbf{r} \, (3x_m \, x_n - \mathbf{r}^2 \delta_{mn}) \rho \, . \tag{8.6}$$

The result is quite natural for the following reasons.

If the multipole expansion of the vector-type electromagnetic radiation, i.e. of the field in the wave zone, starts with the dipole term, one could expect from the very beginning that for the tensor-type gravitational interaction such an expansion should start with the quadrupole.

On the other hand, it is well known that for a system of particles with the same ratio e/m there is neither electric dipole radiation, nor magnetic dipole[2]. And for the gravitational field the role of e/m is played by the ratio of the gravitational mass to the inert one, which according to the principle of equivalence, is the same for all particles. Therefore, there should be no dipole gravitational radiation.

Now we go over to the calculation of the gravitational radiation intensity. To this end we need the energy flux density of the gravitational field, i.e. t_0^n component of its energy-momentum tensor (EMT). As to the EMT of matter T_ν^μ, it satisfies the condition

$$T_{\nu;\mu}^\mu = T_{\nu,\mu}^\mu + \Gamma_{\rho\mu}^\mu T_\nu^\rho - \Gamma_{\mu\nu}^\rho T_\rho^\mu = \frac{1}{\sqrt{-g}} \partial_\mu (\sqrt{-g} \, T_\nu^\mu) - \Gamma_{\mu\nu}^\rho T_\rho^\mu = 0 \, .$$

Due to the term $-\Gamma_{\mu\nu}^\rho T_\rho^\mu$ in this relation, the matter EMT T_ν^μ is not conserved, which is quite natural in the presence of the gravitational field. But then one

---

[2]See, for instance, L.D. Landau and E.M. Lifshitz, *The Classical Theory of Fields*, §67,71.

should build from $g_{\mu\nu}$ (or $h_{\mu\nu}$) such a structure t_ν^μ that would guarantee the relation

$$\frac{1}{\sqrt{-g}}\,\partial_\mu[\sqrt{-g}\,(T_\nu^\mu + t_\nu^\mu)] = 0\,.$$

Then we obtain in the standard way the conservation law for the 4-momentum:

$$P_\nu = \int d\mathbf{r}\sqrt{-g}\,(T_\nu^0 + t_\nu^0) = \text{const}\,. \tag{8.7}$$

However, there is no true tensor t_ν^μ. Indeed, in virtue of the equivalence principle, one can always choose a reference frame in such a way that at any given point the metric will be flat and its first derivatives will turn to zero. But then the structure t_ν^μ, which is built from the first derivatives, turns to zero as well. For a true tensor it means that the tensor vanishes identically. Still, a corresponding *pseudotensor* t_ν^μ, which behaves as a tensor under linear coordinate transformations, can be constructed, and even not in a unique way. For an asymptotically flat system the corresponding total energy and momentum are conserved and defined uniquely by the integrals (8.7).

In the case of interest to us, that of a weak gravitational wave, the construction of the pseudotensor t_ν^μ is sufficiently simple. Let us start with the action (6.8). We note first of all that in our case the term

$$-\sqrt{-g}g^{\mu\nu}\Gamma_{\mu\nu}^\rho\,\Gamma_{\rho\tau}^\tau = \partial_\mu(\sqrt{-g}g^{\rho\mu})\Gamma_{\rho\tau}^\tau$$

in the action turns to zero in accordance with the harmonic condition (6.15) and therefore can be immediately omitted. By the way, it can be easily demonstrated that the second factor in this term, $\Gamma_{\rho\tau}^\tau$, also vanishes for the gravitational wave. Other terms of second order in $h_{\mu\rho}$ in the action, after integrating by parts and omitting total derivatives, reduce to

$$S_g^{(2)} = \frac{1}{64\pi k}\int d^4x\,h_{\mu\nu,\,\rho}h_{\mu\nu,\,\rho}\,.$$

We have taken into account here the harmonic condition and the fact that $h_{\mu\mu} = 0$. Recalling also that $h_{0\mu} = 0$ in the wave, we present the integrand, which is the Lagrangean density for the free gravitational wave, as follows:

$$L_g^{(2)} = \frac{1}{64\pi k}\,h_{mn,\,\rho}h_{mn,\,\rho}\,. \tag{8.8}$$

Now the energy flux density is calculated in the usual way:[3]

$$t_0^3 = h_{mn,\,0}\frac{\partial L_g^{(2)}}{\partial h_{mn,\,3}} = \frac{1}{32\pi k}\,\dot{h}_{mn}\dot{h}_{mn}\,. \tag{8.9}$$

We have taken into account here that in a plane wave $h_{mn,\,3} = -h_{mn,\,0}$.

---

[3]See, for instance, L.D. Landau and E.M. Lifshitz, *The Classical Theory of Fields*, §32.

In a plane wave propagating along the axis 3 the contraction $h_{mn}h_{mn}$ reduces to $h_{11}^2 + h_{22}^2 + 2h_{12}^2$. However, the calculation of the total intensity requires integrating over the angles, i.e. over the directions $\mathbf{n}$ of the wave propagation. Therefore, the expression $h_{11}^2 + h_{22}^2 + 2h_{12}^2$ should be rewritten in the form valid for an arbitrary $\mathbf{n}$, but not only for that directed along the axis 3. First, the tensor, the square of which enters the result, should belong to the plane orthogonal to $\mathbf{n}$, i.e. it should satisfy the condition $n_m h_{mn} = 0$. Such transverse tensor is

$$h_{mn}^\perp = h_{mn} - n_m n_i\, h_{in} - n_n n_i\, h_{mi} + n_m n_n n_i n_j\, h_{ij}.$$

But still this is not all. The needed tensor should also be traceless. Therefore, its correct form for an arbitrary direction $\mathbf{n}$ is

$$\bar{h}_{mn} = h_{mn}^\perp - \frac{1}{2}\left(\delta_{mn} - n_m n_n\right) h_{ii}^\perp.$$

Thus, the contraction entering the answer, for an arbitrary $\mathbf{n}$ equals $\bar{h}_{mn}\,\bar{h}_{mn}$. Simple transformations give

$$\bar{h}_{mn}\,\bar{h}_{mn} = h_{mn}\, h_{mn} - 2n_i n_j\, h_{mi}\, h_{jn} + \frac{1}{2}\left(n_m n_n\, h_{mn}\right)^2.$$

It is convenient to average at once this expression over the directions of $\mathbf{n}$. By means of formulae

$$\langle n_i n_j \rangle = \frac{1}{3}\,\delta_{ij}, \quad \langle n_i n_j n_m n_n \rangle = \frac{1}{15}\left(\delta_{ij}\,\delta_{mn} + \delta_{im}\,\delta_{jn} + \delta_{in}\,\delta_{jm}\right), \quad (8.10)$$

we obtain

$$\langle \bar{h}_{mn}\,\bar{h}_{mn} \rangle = \frac{2}{5}\,h_{mn}\, h_{mn}. \tag{8.11}$$

Now, with relations (8.6), (8.9), and (8.11), we obtain the final result for the total intensity of the gravitational quadrupole radiation (*A. Einstein,* 1918; *M. von Laue,* 1918):

$$I = 4\pi R^2\,\frac{1}{32\pi k}\,\langle \bar{h}_{mn}\,\bar{h}_{mn} \rangle = \frac{k}{45c^5}\,\dddot{q}_{mn}\dddot{q}_{mn}. \tag{8.12}$$

We have restored in the last expression the explicit dependence on the velocity of light c.

As it should be expected, the obtained result (8.12) is very close in structure to the corresponding formula for the electromagnetic quadrupole radiation.[4] In particular, it is also of fifth order in $1/c$.

The discussed effect is extremely small, the registration of gravitational radiation from any conceivable source on the Earth is absolutely unrealistic.

---

[4]See, for instance, L.D. Landau and E.M. Lifshitz, *The Classical Theory of Fields,* §71.

Problem

8.4. Derive relations (8.10).

8.3 Gravitational Radiation of Binary Stars

As to the detection of gravitational waves from cosmic sources, and from binary stars in particular, the situation is different. Let us consider, therefore, in more detail the problem of gravitational radiation of two bodies bound by gravitational interaction. If the distance between the bodies is much larger than the size of both, these bodies can be considered as point-like. Then

$$\rho(\mathbf{r}) = m_1 \, \delta(\mathbf{r} - \mathbf{r}_1(t)) + m_2 \, \delta(\mathbf{r} - \mathbf{r}_2(t)) \, ;$$

here $m_{1,2}$ are the masses of the bodies, $\mathbf{r}_{1,2}(t)$ are their trajectories. The quadrupole moment of the system is

$$q_{mn} = \mu \left(3 r_m r_n - \mathbf{r}^2 \delta_{mn}\right),$$

where $\mu = m_1 m_2/(m_1 + m_2)$ is the reduced mass, and $\mathbf{r} = \mathbf{r}_1(t) - \mathbf{r}_2(t)$ is the relative coordinate. Using the equation of motion

$$\ddot{\mathbf{r}} = -\frac{km}{r^3} \mathbf{r}, \quad m = m_1 + m_2 \, ,$$

and the result of its differentiation over time,

$$\dddot{\mathbf{r}} = -\frac{km}{r^3} \left[\mathbf{v} - \frac{3\mathbf{r}(\mathbf{rv})}{\mathbf{r}^2} \right], \quad \mathbf{v} = \dot{\mathbf{r}} \, ,$$

we obtain

$$\dddot{q}_{mn} \dddot{q}_{mn} = 24 \left(\frac{km_1 m_2}{r^3} \right)^2 \left[12 r^2 v^2 - 11(\mathbf{rv})^2 \right]$$

$$= 24 \left(\frac{km_1 m_2}{r^3} \right)^2 \left[r^2 v^2 + 11 \frac{\mathbf{l}^2}{\mu^2} \right], \tag{8.13}$$

where $\mathbf{l} = \mu \left[\mathbf{r} \times \mathbf{v} \right]$ is the orbital angular momentum of the system.

In the simple case of a circular orbit (when $(\mathbf{rv}) = 0$), we get easily the total intensity of radiation, i.e. the energy loss in unit time:

$$I = -\frac{dE}{dt} = \frac{32}{5} \frac{k^4 m_1^2 m_2^2 (m_1 + m_2)}{c^5 r^5} \, . \tag{8.14}$$

This energy loss results in particular in a decrease in orbital period T. This decrease can be expressed via the world constants k and c, and via masses m_1, m_2, and period T itself, which are measurable. The result for a circular orbit is

$$\dot{T} = -\frac{192\,\pi\,k^{5/3}}{5\,c^5} \left(\frac{T}{2\pi}\right)^{-5/3} m_1 m_2 \,(m_1 + m_2)^{-1/3}. \qquad (8.15)$$

The gravitational radiation even of binary stars has not been directly measured up to now. There is, however, convincing quantitative proof that it indeed exists. The detailed measurements of the pulses of radio waves from the binary pulsar B1913+16 (see short information on it in section 6.4) have demonstrated that the orbital period of this binary star decreases with the rate $-(2.4056 \pm 0.0051) \times 10^{-12}$ s/s. The effect is exactly the same quantitatively as it should be due to the energy loss caused by the gravitational radiation. The ratio of the measured rate T_m to the calculated one T_c (of course, with the account for the orbit eccentricity, see problem 8.7) is

$$\frac{T_m}{T_c} = 1.0013 \pm 0.0021.$$

Let us note that the energy of gravitational waves is huge in the present case, it is quite comparable to the total energy of the Sun's radiation.

It is expected that in the next few years the gravitational radiation of binary stars will be directly registrated by detectors using laser interferometers.

Problems

8.5. Derive relation (8.13).

8.6. Derive formula (8.15).

8.7. Find the relation between the energy loss and the change of the orbital angular momentum, caused by the gravitational radiation, for the circular orbits of components of a binary star.

8.8. Find the intensity of the gravitational radiation, averaged over the rotation period, for the elliptic orbits of components of a binary star (*P.C. Peters, J. Matthews*, 1963).

8.9. Find the change of the orbital angular momentum, averaged over the rotation period, for the elliptic orbits of components of a binary star (*P.C. Peters*, 1964).

8.10. Find the change of the eccentricity of elliptic orbits of the components of a binary star (*P.C. Peters*, 1964).

8.11. Find the change of rotation period for elliptic orbits of the components of a binary star (*P.C. Peters*, 1964).

8.12. A particle with the velocity $v_\infty \ll 1$ at infinity and the impact parameter $\rho = 2r_g(1 + \delta)/v_\infty$, $\delta \ll 1$ scatters on a black hole (see problem 6.14). Estimate the total energy loss due to the gravitational radiation if the particle goes to infinity again (*Ya.B. Zel'dovich, I.D. Novikov,* 1964). How does the total cross-section change due to the gravitational radiation?

8.13. Let us assume that there is a massless scalar field interacting with the energy-momentum tensor of the usual matter. Estimate the intensity of the radiation of this scalar field by a binary star.

8.14. Estimate the total energy loss due to the gravitational radiation when two bodies of comparable masses scatter in such a way that the minimum distance is on the order of their gravitational radii.

8.4 Resonance Transformation of Electromagnetic Wave to Gravitational One

Let a free wave with the electric and magnetic field strengths **e** and **b**, respectively, propagate in a constant external field with field strengths **E** and **B**. Then it is the total field strengths that contribute to the stress tensor T_{mn}, which is the source of the gravitational wave h_{mn}, so that

$$-\Box h_{mn} = 16\pi k T_{mn} = 4k[(E + e)_m(E + e)_n + (B + b)_m(B + b)_n]. \quad (8.16)$$

In the present case it is just T_{mn} (but not $T_{mn} + 1/2\,\delta_{mn}T_\lambda^\lambda$), which enters the right-hand side of the wave equation, since for the electromagnetic field $T_\lambda^\lambda = 0$.

Obviously, the constant part of T_{mn} does not generate the gravitational wave, so that $E_m E_n + B_m B_n$ in the right-hand side can be at once omitted. Then, if the electromagnetic field is weak, one can certainly neglect as well the contribution of $e_m e_n + b_m b_n$ into T_{mn}.

In fact, even if the free wave is strong, its stress tensor

$$\tau_{mn} \sim e_m e_n + b_m b_n$$

in principle cannot generate a gravitational wave. To prove it, let us choose the direction **n** of the wave propagation as the z axis. Since in a free electromagnetic wave $\mathbf{b} = \mathbf{n} \times \mathbf{e}$, in our case $b_1 = -e_2$ and $b_2 = e_1$. Therefore, $\tau_{11} = \tau_{22} \sim e_1 e_1 + e_2 e_2$, and $\tau_{12} = \tau_{21} = 0$. Obviously, such a stress tensor cannot serve as a source of a gravitational wave which should also propagate along the z axis: in this gravitational wave at least one of two polarizations should be distinct from zero, either $h_{11} = -h_{22}$, or $h_{12} = h_{21}$.

Thus, it is sufficient to keep in the right-hand side of (8.16) only the interference terms $E_m e_n + E_n e_m + B_m b_n + B_n b_m$.

The characteristic features of the discussed phenomenon can be elucidated with the following specific example. Let the external field be purely magnetic and directed along the x axis: $\mathbf{B} = (B, 0, 0)$, and the magnetic field of the wave be as follows: $\mathbf{b} = (0, be^{i\omega(z-t)}, 0)$. Then equation (8.16) reduces to

$$(-\partial_0^2 + \partial_z^2)h_{12} = 4kBbe^{i\omega(z-t)}.$$

Its solution is

$$h_{12} = -2ikBb\,\frac{z}{\omega}\,e^{i\omega(z-t)}. \tag{8.17}$$

The energy flux in the gravitational wave is

$$t_0^3 = \frac{1}{32\pi k}\,(\mathrm{Re}\,\dot{h}_{12})^2 = \frac{kB^2b^2z^2}{8\pi}\,\cos^2\omega(z-t).$$

We are interested here only in the contribution to t_0^3 that grows with z as z^2. Therefore, as well as in the case of a common plane wave, we differentiate over z only the exponential in expression (8.17). As to the energy flux in the electromagnetic wave, it is obviously

$$T_0^3 = \frac{b^2}{4\pi}\,\cos^2\omega(z-t).$$

Thus, a resonance transition of an electromagnetic wave into the gravitational one takes place in an external field (*M.E. Gertsenstein*, 1961), with the transformation factor

$$K = \frac{t_0^3}{T_0^3} = \frac{1}{2}\,kB^2z^2.$$

Despite the resonant character of the transition, i.e. the linear growth with z of the gravitational wave amplitude, the effect is so weak that one can hardly hope to observe it in the conceivable future.

Still, the discussion of this phenomenon is not only of methodological interest. Searches for other, nongravitational, hypothetical fields with zero or very small rest mass are based on analogous effects.

Problem

8.15. How does the quadratic growth of the energy flux of gravitational wave change at very large distances?

8.5 Synchrotron Radiation of Ultrarelativistic Particles Without Special Functions

The synchrotron radiation, i.e. the radiation of a charged particle in an external magnetic field, is considered in numerous textbooks.[5] However, the

---

[5]See, for instance, L.D. Landau and E.M. Lifshitz, *The Classical Theory of Fields*, §74.

consideration is based usually on the analysis of the exact solution of the problem. For the radiation of relativistic particles in a gravitational field such an analysis and exact solution by itself are incomparably more complicated. In the case of radiation in a gravitational field the qualitative analysis is not only more transparent, more instructive from the physical point of view, but certainly more practical also.

The detailed qualitative analysis of the common synchrotron radiation performed below serves as an introduction to the next section where the radiation of ultrarelativistic particles in a gravitational field is considered. One may think that the arguments presented here will be useful by themselves, irrespective to the problems considered in the next section.

Let us start with the total radiation intensity. In the locally inertial frame (LIF) comoving with an electron, it is

$$I' \sim e^2 (a')^2 \sim \frac{e^4}{m^2} (E')^2. \tag{8.18}$$

Here e and m are the electron charge and mass, a is its acceleration, E is the electric field strength; I, a and E are supplied with primes to point out that they refer to the LIF. E' is obtained from the magnetic field B in the laboratory frame (LF) by the Lorentz transformation

$$E' \sim B\gamma, \quad \gamma = \frac{1}{\sqrt{1 - v^2}}. \tag{8.19}$$

We recall now that I is an invariant. Indeed, the radiation intensity is expressed through the probability of the photon emission W and its energy $\hbar\omega$ as follows: $I = W\hbar\omega$. Then, the probability W in the LF is related to the probability in the LIF W' by the relation $W = W'/\gamma$ (just recall that the lifetime of an unstable particle in LF is γ times larger than that in LIF). On the other hand, it is well known that $\omega = \omega'\gamma$. Finally, $I' = I$.

Now, substituting into (8.18) expression (8.19) for the electric field E' in the LIF, we obtain the well-known result

$$I \sim \frac{e^4}{m^2} B^2\gamma^2. \tag{8.20}$$

If instead of B one fixes the radius of the electron trajectory r_0, related to B via $eB \sim m\gamma/r_0$, the expression for the total intensity becomes

$$I \sim \frac{e^2\gamma^4}{r_0^2}. \tag{8.21}$$

Let us go over now to the angular distribution of the radiation. In the LIF it has a common dipole form, it is just trigonometry. In other words, in LIF $\theta' = k_t'/k_l' \sim 1$. Here $k_{t(l)}'$ is the transverse (longitudinal) component of the wave vector of the photon. In the LF these components are: $k_t = k_t'$, $k_l = k_l'\gamma$.

Therefore, in the LF an ultrarelativistic electron radiates into a cone with a typical angle

$$\theta_c \sim k_t/k_l \sim \gamma^{-1}. \tag{8.22}$$

An observer receives the signal only staying inside this cone which moves together with the electron. An elementary consideration demonstrates that the electron beams at the observer only from the piece of the trajectory arc that has the same angular size as the cone itself. In the present case it means that the angular size of this piece of the arc is $\theta_c \sim \gamma^{-1}$. In other words, the formation length for radiation, which in our ultrarelativistic case ($v \approx c = 1$) coincides with the formation time for it, is

$$\Delta t \sim r_0 \theta_c \sim r_0 \gamma^{-1}.$$

Then the duration of signal receiving, with the account for the longitudinal Doppler effect, is

$$\delta t = (1 - \mathbf{n}\mathbf{v})\Delta t \approx \frac{1}{2}(\theta^2 + \gamma^{-2})\Delta t, \tag{8.23}$$

where $\mathbf{n} = \mathbf{k}/k$. For $\theta \sim \theta_c \sim \gamma^{-1}$ we obtain $\delta t_c \sim r_0 \gamma^{-3}$. It means that the characteristic frequency of the received radiation is γ^3 times larger than the rotation frequency ω_0:

$$\omega_c \sim \delta t_c^{-1} \sim \gamma^3 r_0^{-1} \sim \gamma^3 \omega_0. \tag{8.24}$$

We turn now to the spectral distribution of the synchrotron radiation. Its intensity decreases rapidly for $\omega \gg \omega_c$. Let us assume that for $\omega \lesssim \omega_c$ it changes according to a power law: $I(\omega) \sim \omega^\nu$. Then, by comparing the total intensity given by the integral

$$\int^{\omega_c} d\omega I(\omega) \sim \omega_c^{\nu+1} \sim \gamma^{3(\nu+1)}$$

with formula (8.21), we obtain $\nu = 1/3$. In other words,

$$I(\omega) \sim \omega^{1/3} \quad \text{for} \quad \omega \lesssim \omega_c, \tag{8.25}$$

or for the discrete spectrum

$$I_n \sim n^{1/3} \quad \text{for} \quad n \lesssim \gamma^3. \tag{8.26}$$

And at last, let us find the angular distribution of the radiation for the frequency range

$$\omega_0 \ll \omega \ll \omega_c, \qquad 1 \ll n \ll \gamma^3.$$

It is natural to expect that here the characteristic angles θ are larger than γ^{-1}. As previously, while the angle of the radiation cone is small, $\theta \ll 1$, the electron beams at the observer only from the piece of the trajectory arc which has the same angular size θ. But then, instead of relation (8.23), we obtain

$$\delta t \sim \omega^{-1} \sim \theta^2 \Delta t \sim \theta^3 r_0 \sim \theta^3 \omega_0^{-1}.$$

Thus, in this frequency region

$$\theta \sim \left(\frac{\omega_0}{\omega}\right)^{1/3} \sim n^{-1/3}. \tag{8.27}$$

In the conclusion of this section, it should be emphasized that the obtained qualitative results are not special for the considered problem of finite motion of an ultrarelativistic particle in a magnetic field. They are applicable as well to a more general case, that of scattering in external electromagnetic fields if characteristic scattering angles exceed γ^{-1}.

Problem

8.16. An ultrarelativistic electron scatters by a large angle in an external electromagnetic field. Find the momentary intensity of the gravitational radiation. In the present case the basic mechanism of its generation is the resonance transformation of the electromagnetic synchrotron radiation into the gravitational one (*I.B. Khriplovich, O.P. Sushkov, 1974*).

8.6 Radiation of Ultrarelativistic Particles in Gravitational Field

In this section, as well as in the previous one, we are not confined to the case of a circular motion, that for the Schwarzschild field is of methodological interest only, due to instability of ultrarelativistic circular orbits. We discuss also the radiation under infinite motion. The presented approach is due to *I.B. Khriplovich, E.V. Shuryak* (1973).

It can be easily seen that in this case as well, the radiation of a particle is concentrated in the region of angles $\theta \sim 1/\gamma$ (see (8.22)). Still, there is a serious distinction from the radiation in an external electromagnetic field. It is as follows. In an external gravitational field, just in virtue of the equivalence principle, the trajectory of an emitted particle, photon or graviton, is very close to the trajectory of its ultrarelativistic emitter. Thus, here the formation length of the radiation, both electromagnetic and gravitational, for the circular motion coincides in the order of magnitude with the radius of the trajectory r_0, but is not contracted as compared to it by γ times, as was the case in an external electromagnetic field. Due to the nonlocal formation of the radiation, it does not make sense in the present case to talk about its total intensity in the LIF.

Therefore, we will estimate the differential intensity dI in the element $d\Omega$ of solid angle with the standard formula

$$dI \sim \omega^2 u^2 R^2 d\Omega \, \frac{\partial t}{\partial t'} \, .$$

Here u is the characteristic amplitude of the field of the wave; $\omega^2 u^2$ is the estimate for the energy flux density, i.e. for T_0^3 and t_0^3 in the cases of electromagnetic and gravitational waves, respectively. The factor $\partial t / \partial t'$ is due to the fact that the intensity is being measured with respect to the time t of the observer, but not to the time t' of the emitter. Since $t = t' + |\mathbf{r} - \mathbf{r}(t')|$, for an ultrarelativistic particle we have

$$\frac{\partial t}{\partial t'} = 1 - \mathbf{v}(t') \cdot \frac{\mathbf{r} - \mathbf{r}(t')}{|\mathbf{r} - \mathbf{r}(t')|} = 1 - \mathbf{nv} \approx \frac{1}{2} \left(\frac{1}{\gamma^2} + \theta^2 \right),$$

or

$$\Delta t \sim \left(\frac{1}{\gamma^2} + \theta^2 \right) \Delta t'. \tag{8.28}$$

With the radiation concentrated in the angular interval $\theta \sim 1/\gamma$, its characteristic frequencies are

$$\omega_c \sim \gamma^2 \omega_0 \sim \gamma^2 \frac{1}{r_0} \, . \tag{8.29}$$

They exceed the rotation frequency not by γ^3 times, as was the case with the synchrotron radiation (see (8.24)), but only by γ^2 times. In this respect the situation here resembles that taking place for a scattering by a small angle, less than $1/\gamma$, in an external electromagnetic field.

For $\omega \lesssim \omega_c$ the radiation propagates inside characteristic angles $\theta \gtrsim 1/\gamma$ with respect to the direction of motion of the emitter, so that

$$\omega \sim \frac{1}{\Delta t} \sim \frac{1}{r_0 \theta^2} \, ,$$

or

$$\theta \sim (\omega r_0)^{-1/2} \sim \left(\frac{\omega_0}{\omega} \right)^{1/2} \tag{8.30}$$

(but not $(\omega_0/\omega)^{1/3}$, as was the case with the synchrotron radiation, see (8.27)).

Let us turn now from these general relations to concrete expressions for the electromagnetic and gravitational radiation. In the formulae below they are described by the first and second expression, respectively.

The three-dimensionally transverse (with respect to $\mathbf{n}$) field amplitudes u in the wave zone are:

$$A_\perp \sim \frac{ev_\perp}{1 - \mathbf{nv}} \sim \frac{e\theta}{\theta^2 + 1/\gamma^2} \, , \quad h_{\perp\perp} \sim \frac{\sqrt{k}\varepsilon(v_\perp)^2}{1 - \mathbf{nv}} \sim \frac{\sqrt{k}\varepsilon\theta^2}{\theta^2 + 1/\gamma^2} \, .$$

We recall here that $\sqrt{k}\varepsilon$, where ε is the particle energy, plays the same role in gravity as e in electrodynamics.

The differential intensities of radiation for the angles $\theta \gtrsim 1/\gamma$ are

$$\frac{dI_{em}}{d\theta} \sim \frac{e^2}{r_0^2} \frac{1}{\theta^3}, \quad \frac{dI_{gr}}{d\theta} \sim \frac{k\varepsilon^2}{r_0^2} \frac{1}{\theta} = \frac{km^2\gamma^2}{r_0^2} \frac{1}{\theta}.$$

And at last, the total intensities are

$$I_{em} \sim \frac{e^2}{r_0^2} \int_{1/\gamma}^1 \frac{d\theta}{\theta^3} \sim \frac{e^2\gamma^2}{r_0^2}, \quad I_{gr} \sim \frac{k\varepsilon^2}{r_0^2} \int_{1/\gamma}^1 \frac{d\theta}{\theta} \sim \frac{k\varepsilon^2}{r_0^2} \ln\gamma = \frac{km^2\gamma^2}{r_0^2} \ln\gamma.$$

The corresponding frequency spectra for $\omega \lesssim \omega_c \sim \gamma^2/r_0$ are

$$I_{em} \sim \omega^0 = \text{const}, \quad I_{gr} \sim \ln\omega. \tag{8.31}$$

Now we briefly discuss a more realistic problem, that of the radiation of ultrarelativistic particles moving with the impact parameter ρ in the Schwarzschild field. In this case the duration of the signal is $\Delta t \sim \rho/\gamma^2$, so that the characteristic frequencies are $\omega_c \sim \gamma^2/\rho$. The total intensity of radiation can be obtained from the corresponding formulae for the circular motion by the substitution $1/r_0^2 \to r_g^2/\rho^4$. Indeed, while for the circular motion the acceleration equals $dv/dt' \sim 1/r_0$, in the scattering problem it is $dv/dt' \sim r_g/\rho^2$. Then, we multiply the intensity by the characteristic time of flight, and obtain in this way the following estimates for the total energy loss:

$$\Delta\varepsilon_{em} \sim \frac{e^2 r_g^2 \gamma^2}{\rho^3}, \quad \Delta\varepsilon_{gr} \sim \frac{k\varepsilon^2 r_g^2}{\rho^3} \ln\gamma = \frac{km^2\gamma^2 r_g^2}{\rho^3} \ln\gamma.$$

Problems

8.17. Derive relations (8.31).

8.18. An ultrarelativistic particle with the impact parameter

$$\rho = (3\sqrt{3}/2)r_g(1+\delta), \quad \delta \ll 1,$$

scatters on a black hole (see problem 6.16). Estimate the total energy loss due to the gravitational radiation if the particle goes to infinity again. How does the total cross-section change due to the gravitational radiation?

9

General Relativity and Cosmology

9.1 Geometry of Isotropic Space

The modern cosmology is based on the solution of the Einstein equations found by *A.A. Friedmann* (1922). This solution, in its turn, is based on the assumption that the distribution of matter in the Universe is homogeneous and isotropic.

In the real world, the matter (or at least a large part of the matter) is condensed into stars, stars are condensed into galaxies, and galaxies are condensed into clusters. But on this last stage the inhomogeneities are apparently over: astronomic observations at least are not in conflict with the assumption that the "gas" of the clusters of galaxies is homogeneous and isotropic.

If an n-dimensional space is completely isotropic, its Riemann tensor R_{ijkl} should be characterized by a scalar, namely by the scalar curvature R. Therefore, with the account for the symmetry properties (3.46) and (3.47), the curvature tensor in the locally Euclidean space should appear as follows:

$$R_{ijkl} = K(\delta_{ik}\delta_{jl} - \delta_{il}\delta_{jk}),$$

with the coefficient K being proportional to R. The natural generalization of this equality for arbitrary coordinates is

$$R_{ijkl} = K\left(g_{ik}g_{jl} - g_{il}g_{jk}\right). \qquad (9.1)$$

The coefficient K in this formula is independent of coordinates. This can be easily proven by plugging relation (9.1) into the contracted Bianchi identity (3.53). Thus, an isotropic space is simultaneously a homogeneous one. However, the constant K may depend on time.

In a three-dimensional space, contracting formula (9.1) in ik and in jl relates the coefficient K to the scalar curvature R as follows:

$$R = 6K.$$

Depending on the sign of the scalar curvature, three essentially different cases are possible for the space metric of an isotropic space: 1) constant positive curvature, $K > 0$; 2) constant negative curvature, $K < 0$; 3) zero curvature, $K = 0$. Clearly, the last case corresponds to the flat, Euclidean space.

It is convenient to investigate the geometry of a space of constant positive curvature by treating it as the geometry on a three-dimensional hypersphere in some auxiliary four-dimensional Euclidean space (of course, unrelated to the four-dimensional space-time). The equation for a hypersphere of radius a in this space is

$$x_1^2 + x_2^2 + x_3^2 + x_4^2 = a^2,$$

and the element of length on it is

$$dl^2 = dx_1^2 + dx_2^2 + dx_3^2 + dx_4^2.$$

By expressing the auxiliary coordinate x_4 via the physical ones x_1, x_2, x_3, and eliminating dx_4^2 from dl^2, we find

$$dl^2 = dx_1^2 + dx_2^2 + dx_3^2 + \frac{(x_1 dx_1 + x_2 dx_2 + x_3 dx_3)^2}{a^2 - x_1^2 - x_2^2 - x_3^2}. \tag{9.2}$$

To relate the constants a^2 and K, we put $x_3 = 0$. It is clear that the surface obtained in this way is a two-dimensional sphere with the Gauss curvature

$$K = \frac{1}{a^2}. \tag{9.3}$$

Now we go from x_1, x_2, x_3 to the spherical coordinates r, θ, ϕ. Instead of direct calculation, one can note that under the shift along the radius, i.e. for $d\mathbf{r} \| \mathbf{r}$, the longitudinal interval is

$$dl_\|^2 = dr^2 \left(1 + \frac{r^2}{a^2 - r^2}\right) = \frac{dr^2}{1 - r^2/a^2}.$$

On the other hand, for the shift $d\mathbf{r} \perp \mathbf{r}$ the transverse interval is $dl_\perp^2 = d\mathbf{r}^2$. Then it is clear that in the spherical coordinates

$$dl^2 = \frac{dr^2}{1 - r^2/a^2} + r^2 (d\theta^2 + \sin^2 \theta \, d\phi^2). \tag{9.4}$$

Of course, any point of the space can be chosen as the origin. The length of a circle in these coordinates is $2\pi r$, and the surface area of a sphere is $4\pi r^2$. The length of the radius of a circle and sphere

$$\int_0^r \frac{dr}{\sqrt{1 - r^2/a^2}} = a \arcsin \frac{r}{a}$$

exceeds r.

Then, it is convenient to introduce four-dimensional spherical coordinates

$$a, \quad 0 \leq \chi \leq \pi, \quad 0 \leq \theta \leq \pi, \quad 0 \leq \phi \leq 2\pi$$

in the auxiliary space, so that

$$x_1 = a \sin \chi \sin \theta \cos \phi, \quad x_2 = a \sin \chi \sin \theta \sin \phi, \quad x_3 = a \sin \chi \cos \theta,$$

$$x_4 = a \cos \chi.$$

Obviously, now $r = a \sin \chi$ and the interval becomes

$$dl^2 = a^2 [d\chi^2 + \sin^2 \chi (d\theta^2 + \sin^2 \theta d\phi^2)]. \tag{9.5}$$

In the new variables the distance of a point from the origin is $a\chi$. With the increase of this distance, the surface area of a sphere $S = 4\pi a^2 \sin^2 \chi$ at first increases and reaches at the distance $\pi a/2$ its maximum value equal to $4\pi a^2$. Then it starts to decrease and turns to zero at the maximum possible distance πa. The volume of a four-dimensional space with positive curvature is finite:

$$V = \int_0^{2\pi} d\phi \int_0^{\pi} \sin \theta \, d\theta \int_0^{\pi} \sin^2 \chi \, d\chi \, a^3 = 2\pi^2 a^3. \tag{9.6}$$

However, this space has no boundaries. Hence it follows in particular that the total electric charge in such a space should be equal to zero. Indeed, any closed surface in a finite space splits this space into two finite domains. The flux of electric field through this surface is equal to the total charge of a domain situated on one side of the surface. But the same flux is equal to the total charge of another domain situated on the opposite side of the surface taken with the opposite sign. It is clear that the sum of the charges from both sides of the surface should be equal to zero. By the analogous reason, the total 4-momentum of a closed space should also vanish.

Let us discuss the spaces of constant negative curvature. It follows from (9.3) that formally this corresponds to the substitution $a \to ia$. Therefore, the geometry of a space of constant negative curvature corresponds to the geometry on a four-dimensional pseudosphere of imaginary radius. Now

$$K = -\frac{1}{a^2},$$

and the element of length in the coordinates r, θ, ϕ is

$$dl^2 = \frac{dr^2}{1 + r^2/a^2} + r^2(d\theta^2 + \sin^2 \theta d\phi^2),$$

with $0 \leq r \leq \infty$. The change of variables $r = a \sinh \chi$ gives

$$dl^2 = a^2 [d\chi^2 + \sinh^2 \chi (d\theta^2 + \sin^2 \theta \, d\phi^2)]. \tag{9.7}$$

The volume of a space of negative curvature is infinite.

Of course, the case of a flat, Euclidean space, with $K = 0$, is also possible.

Problems

9.1. Prove that K in expression (9.1) is independent of coordinates.

9.2. Prove equality (9.3) by direct calculation of the scalar curvature of space. The calculation is conveniently performed in the vicinity of the origin, i.e. for small x_1, x_2, x_3.

9.3. Transform interval (9.4) to the form where it is proportional to the Euclidean expression.

9.4. Prove relation (9.6).

9.2 Isotropic Model of the Universe

In the case of a closed Universe, the visual two-dimensional analogue of the solution we are looking for is an inflating sphere, a soap bubble. In the comoving reference frame the matter on the sphere is at rest, i.e. the angular coordinates of each particle of the dust do not change, and only the radius of the sphere grows with time. In our problem of a three-dimensional space the coordinates χ, θ, ϕ of each cluster of galaxies remain constant, only the scale of the distances $a(t)$ grows.

Since there is no singled out direction in the space, the components g_{0m} ($m = 1, 2, 3$) of the metric tensor, which constitute a three-dimensional vector, should vanish. The component g_{00} depends only on t, so that by a suitable choice of the time coordinate one can turn $g_{00}dt^2$ into dt^2. Thus, the four-dimensional interval transforms to

$$ds^2 = dt^2 - dl^2 = dt^2 - a^2(t)[d\chi^2 + \sin^2\chi(d\theta^2 + \sin^2\theta d\phi^2)].$$

It is convenient to change from the time t to a new variable η defined by relation $dt = a(t)d\eta$. In result, the interval is written as

$$ds^2 = a^2(\eta)[d\eta^2 - d\chi^2 - \sin^2\chi(d\theta^2 + \sin^2\theta d\phi^2)]. \tag{9.8}$$

To write down the Einstein equations, one should calculate the Ricci tensor. First of all, it is the curvature of the three-dimensional space that contributes to it. This contribution is found at once from formula (9.1) (taking into account that the four-dimensional space is pseudoeuclidean):

$$R_{ij}^{(1)} = -\frac{2}{a^2}\,g_{ij}. \tag{9.9}$$

Another contribution to the Ricci tensor is due to the dependence of the metric on η. The components of the Christoffel symbol with the derivative over η (denoted below by prime) are

$$\Gamma^0_{00} = \frac{a'}{a}, \quad \Gamma^0_{ij} = -\frac{a'}{a^3} g_{ij}, \quad \Gamma^i_{0j} = \frac{a'}{a} \delta^i_j. \tag{9.10}$$

The components Γ^0_{0i} and Γ^i_{00} vanish since there is no singled out direction in our three-dimensional space. By the same reason, turn to zero the components R_{0i} of the Ricci tensor. Its purely time component is

$$R_{00} = -3 \left(\frac{a''}{a} - \frac{a'^2}{a^2} \right). \tag{9.11}$$

At last, the corresponding contribution to the purely space components is

$$R^{(2)}_{ij} = -g_{ij} \left(\frac{a''}{a^3} + \frac{a'^2}{a^4} \right). \tag{9.12}$$

The total expression for the space components of the Ricci tensor is

$$R_{ij} = R^{(1)}_{ij} + R^{(2)}_{ij} = -g_{ij} \left(\frac{a''}{a^3} + \frac{a'^2}{a^4} + \frac{2}{a^2} \right). \tag{9.13}$$

The scalar curvature is

$$R = -6 \left(\frac{a''}{a^3} + \frac{1}{a^2} \right),$$

so that the 00 component of the Einstein equation is written as

$$R_{00} - \frac{1}{2} g_{00} R = 3 \left(\frac{a'^2}{a^2} + 1 \right) = 8\pi k\, T_{00}.$$

Quite analogous calculations in the case of the open Universe result in the equation that differs from this one only by the sign at 1 in the bracket. And for the Universe where the three-dimensional space by itself is flat, Euclidean, the bracket simplifies to a'^2/a^2. As to the right-hand side, since $u^0 = d\eta/ds = 1/a$, we obtain in all three cases, for the closed, open, and flat Universe,

$$T_{00} = g_{00} g_{00} \rho\, u^0 u^0 = \rho a^2.$$

Thus, in the general case of isotropic Universe the discussed equation is

$$\frac{a'^2}{a^4} + \frac{q}{a^2} = \frac{8\pi}{3} k\rho. \tag{9.14}$$

Here and below $q = 1$ for the closed Universe, $q = -1$ for the open Universe, $q = 0$ for the flat Universe.

In the employed comoving reference frame the space components of the four-dimensional velocity are equal to zero, i.e. the coordinates χ, θ, and ϕ of each particle of the dust are independent of η. Therefore, in this frame all components of the energy-momentum tensor, but T_{00}, vanish. The $0n$ components of the Einstein equations turn into the identity $0 = 0$, and their mn components appear as follows:

$$R_{mn} - \frac{1}{2} g_{mn} R = \left(2 \frac{a''}{a^3} - \frac{a'^2}{a^4} + \frac{q}{a^2} \right) g_{mn} = 0 \,,$$

or simply

$$2 \frac{a''}{a^3} - \frac{a'^2}{a^4} + \frac{q}{a^2} = 0 \,. \tag{9.15}$$

We note that while the transition from the three-dimensional geometry of the closed Universe to the geometry of the open one is accompanied by the change $a \to ia$, the corresponding transition in the dynamic equation (9.15) demands one more change: $\eta \to i\eta$.

The volume of the closed Universe grows $\sim a^3$ in the process of expansion, and the total mass of the dust remains constant. So the dust density changes according to the law $\rho = c/a^3$. With this dependence, we arrive for the closed Universe at the equation

$$a'^2 + a^2 = 2a_0 a \,, \quad \text{where} \quad a_0 = \frac{4\pi}{3} kc \,,$$

or

$$a'^2 + (a - a_0)^2 = a_0^2 \,.$$

This is obviously the energy integral for the oscillator with equilibrium at the point $a = a_0$. Under corresponding choice of the initial condition, the solution for $a(\eta)$ is

$$a = a_0 (1 - \cos \eta) \,. \tag{9.16}$$

Since by definition $dt = a d\eta$, then

$$t = a_0 (\eta - \sin \eta). \tag{9.17}$$

Equations (9.16) and (9.17) describe in a parametric form the evolution of the closed Universe. This evolution is of cyclic character: the expansion from the point ($a = 0$ for $t = 0$) to $a_{max} = 2a_0$ changes with contraction to the point, and then everything starts anew.

We now examine the case of the open Universe. Looking at (9.16) and recalling that this transition is linked to the change $\eta \to i\eta$, and also that $a > 0$, it is natural to assume that in this case

$$a = a_0 (\cosh \eta - 1). \tag{9.18}$$

One can check easily that function (9.18) indeed satisfies equation (9.15) (for $q = -1$). Correspondingly, in this case

$$t = a_0 (\sinh \eta - \eta). \tag{9.19}$$

Here the expansion from the point ($a = 0$ for $t = 0$) goes on infinitely. The dust density, defined here by equation (9.14), falls down with the increase of η, the regime of the expansion approaches the free one, so that asymptotically the radius a grows linearly with time.

And at last, in the case of the flat Universe, in line with the trivial solution $a = a_0$, $t = a_0\eta$, there is the nontrivial one:

$$a = a_0\eta^2, \quad t = \frac{a_0}{3}\eta^3, \quad \text{or} \quad a(t) \sim t^{2/3}. \tag{9.20}$$

This solution corresponds to the interval of the form

$$ds^2 = dt^2 - a_1 t^{4/3}(dx^2 + dy^2 + dz^2).$$

In fact, the dependence of $a(t) \sim t^{2/3}$ takes place also in two other cases as well, but for small times only. One can easily check it by considering the corresponding formulae, (9.16) and (9.17), (9.18) and (9.19), in the limit $\eta \to 0$. In this limit

$$\frac{da}{dt} \sim t^{-1/3} \to \infty.$$

We note that for the closed Universe the same regime sets in also under the subsequent contraction into the point.

However, near the singularity where the density ρ turns to infinity, the discussed description is inapplicable. First, the "dust" approximation, used for the description of matter is not valid here. But there is an even deeper reason: we deal here with superstrong fields and therefore need the quantum theory of gravity.

Problems

9.5. Derive relations (9.10) through (9.13) for the closed Universe, as well as analogous formulae for the open Universe.

9.6. Find the asymptotic behavior of density ρ in the open Universe for $t \to \infty$.

9.7. Derive relations (9.20) for the flat Universe.

9.3 Isotropic Model and Observations

Let us come back into the present epoch. We choose the position of an observer as the origin in the isotropic Universe. Then his distance to the galaxy with coordinates χ, θ, ϕ is $l = a\chi$, and due to the expansion of the Universe, this galaxy moves away from the observer with the velocity

$$v = \frac{da}{dt}\,\chi = \frac{1}{a}\frac{da}{dt}\,(a\chi) = Hl.$$

Thus, the model results in the remarkable qualitative prediction: the velocity v with which galaxies move away one from another (at a given moment t) is proportional to the distance l between them. This prediction agrees with the observations of the red shift in the spectra of galaxies that is interpreted as the Doppler effect. The numerical value of the proportionality factor, the so-called Hubble constant, obtained from the modern astronomical observations, is

$$H = \frac{1}{a}\frac{da}{dt} = 73 \pm 3\,\text{km s}^{-1}\,\text{Mpc}^{-1} \tag{9.21}$$

(1 Mpc (Megaparsec) $= 10^6$ parsec, 1 parsec $= 3.26$ light years).

We come back now to equation (9.14). It can be rewritten as follows:

$$H^2 + \frac{q}{a^2} = \frac{8\pi}{3}\,k\,\rho,$$

or

$$\rho = \rho_c + \frac{3q}{8\pi k a^2}. \tag{9.22}$$

We introduce here the so-called critical density

$$\rho_c = \frac{3H^2}{8\pi k} \approx 10 \cdot 10^{-30}\,\text{g/cm}^3;$$

this number for ρ_c corresponds to the value (9.21) of the Hubble constant. It is clear from formula (9.22) that the type of the geometry of the universe is determined by the relation between the true density ρ and the parameter ρ_c that depends on the Hubble constant. If the density exceeds the critical one, the Universe is closed; if the density is less than the critical one, it is open; if the density is equal to the critical one, it is flat.

Usually the dimensionless parameter $\Omega = \rho/\rho_c$ is considered. Modern astronomical data give for the density of the common luminous matter the following result:

$$\Omega_l \approx 0.04.$$

However, the density of some invisible dark matter Ω_d is perhaps much higher, as indicated in particular by the estimates of the masses of galaxy clusters, based on the velocity distribution of separate galaxies. Quantitatively, the total matter density is

$$\Omega_m = \Omega_l + \Omega_d \approx 0.25. \tag{9.23}$$

On the other hand, a global analysis of the modern data of observational astronomy indicates that the Universe is flat, that

$$\Omega_{tot} = 1.02 \pm 0.02. \tag{9.24}$$

The gap ~ 0.77 between (9.24) and (9.23) should be filled in by some unknown form of matter. Moreover, this mysterious form of matter should have negative pressure, $p \approx -\rho$, which is required by the observational data indicating that the expansion of the Universe at the present epoch is accelerated. It is considered as a serious indication that this peculiar "matter", filling in the gap between (9.24) and (9.23), is in fact the nonvanishing cosmological constant Λ (see equation (4.3)).

It should be noted that the required magnitude of Λ is extremely small (typical value of the corresponding density is $\Lambda/8\pi k \sim \rho_c \sim 10^{-29}$ g/cm^3), so that it hardly could be observed anywhere, but in cosmology.

We discuss now the relation between the Hubble constant and the age of the Universe. For the flat world, with $\Omega = 1$, it follows from relation (9.20) that $H(t) = 1/a \, da/dt = 2/3 \, t$. It means that, under the assumption of the flat Universe, its age T (i.e. the time elapsed from the moment when the density was infinite) is related to the present value (9.21) of the Hubble constant as follows:

$$T = \frac{2}{3} \frac{1}{H}. \tag{9.25}$$

Obviously, the obtained relation is valid also for the early stages of expansion in the closed and open Universe, where in both cases $a \sim t^{2/3}$. Even at the late stage of expansion of the open Universe, when the density is so small that a grows linearly with time, the relation between T and H differs from (9.25) by a coefficient only: 1 instead of 2/3. Therefore, since the ratio Ω is at any rate not so far from 1, the estimate (9.25) is apparently quite reasonable. The numerical value of the age of the Universe for $H \approx 70$ km s^{-1} Mpc^{-1}, according to (9.25), is

$$T \approx 13 \cdot 10^9 \text{ years.} \tag{9.26}$$

The age of the Earth, according to the data on the content of radioactive isotopes in the crust of the Earth, is about $4 \cdot 10^9$ years. The estimates of the ages of the oldest galaxy clusters appear as $10 \cdot 10^9$ years. Therefore, the age of the Universe certainly cannot be considerably less than result (9.26).

Problem

9.8. Prove that in an isotropic Universe the relation

$$\frac{d\rho}{dt} + 3H(\rho + p) = 0$$

holds. Here p is the pressure of matter $(T_n^m = -p\,\delta_n^m)$.

10

Are Black Holes Really Black?

10.1 Entropy and Temperature of Black Holes

The classical description of black holes, presented earlier in section 6.6, is incomplete in principle. *J. Wheeler* (1971) was the first to realize it. His line of reasoning looked roughly as follows. Let us take a box filled with the black-body radiation at some temperature T. Obviously it possesses a finite entropy as well. We drop the box into a black hole. Then the entropy of the observable part of the universe will decrease forever. But this is an explicit violation of the second law of thermodynamics! To save the second law we are just obliged to assume that the black hole itself has some entropy (*J. Bekenstein*, 1972) which increases when the box is absorbed. But it is only natural to ascribe some finite temperature as well to a system with a finite entropy. In spite of being so unexpected, this conclusion is quite natural from a somewhat different point of view. A black hole is an ideal absorber, an absolutely black body, for which the temperature is a quite natural property.

Let us try at first to estimate this temperature just by dimensional arguments. We will measure the temperature in the same units as energy, getting rid in this way of the Boltzmann constant in our formulae. By the way, in these units the entropy is dimensionless, so that it cannot be estimated by means of dimensional arguments. A black hole by itself is characterized by the only parameter, its mass M. Besides M, we have also at our disposal two fundamental constants, k and c. One of them, the gravitational constant k, should apparently enter the result by the very meaning of the problem. The obvious combination Mc^2 will not do as temperature: it is too large, and does not contain k. But one cannot construct any other combination of the dimension of energy from M, k, and c. But there is one more fundamental constant at our disposal, the Planck constant $\hbar$. By means of $\hbar$, it is no problem to construct the necessary combination of the dimension of energy: the black hole temperature is on the order of magnitude $\hbar c^3/(kM)$, or (up to a factor of two) $\hbar c/r_g$.

The problem is formally solved, but the natural question arises: what has the quantum of action $\hbar$ to do with our problem, that is completely classical at first sight? To reply to the question, we will consider the box filled with radiation from a somewhat different point of view ($R.$ $Geroch$). Now we lower adiabatically this box to the black hole by means of a rope. The rope is wound on a fly-wheel situated far away from the black hole. The fly-wheel rotates, and its energy may be utilized in principle. We recall that the energy of a body in the gravitational field of a black hole is

$$E = mc^2 \sqrt{1 - r_g/r}.$$

Therefore, at this slow, adiabatic lowering of the box filled with radiation, which has the total mass m, the energy extracted in this way is

$$\Delta E = mc^2 \left(1 - \sqrt{1 - r_g/r}\right).$$

When the box approaches the horizon, we open a lid in its bottom. The radiation escapes into the black hole, and then the box is brought with the rope back into its initial position far away from the star. At first sight, the energy extracted in this way equals the energy of all radiation that was stored in the box. One may think that all this radiation has stuck to the horizon.

However, this is not the case. Due to the uncertainty relation, the size of the box cannot be smaller than the characteristic wave length of radiation λ. In its turn, the characteristic energy of quanta $\hbar\omega$ is nothing but the radiation temperature T_1. Therefore, the size of the box, its height included, is bounded by the condition

$$d > \frac{\hbar c}{T_1}.$$

On the other hand, it is in principle important here to be able to bring the box back to its initial position, far away from the horizon. Therefore, the upper wall of the box certainly stays at a distance, that exceeds d, from the horizon. Then it is natural that not all the radiation contained in the box can be transformed into work, but its part only, limited by relation

$$\eta \sim 1 - \frac{d}{r_g} < 1 - \frac{\hbar c}{r_g T_1}. \tag{10.1}$$

The discussed system, consisting of the black hole and the box, filled with radiation and attached to a rope, can be considered as a heat machine with the working body, which is radiation, of the temperature T_1. Now, η is nothing but the efficiency of this machine, and hence it is well known to be bounded by the Carnot formula

$$\eta_{max} = 1 - \frac{T}{T_1}, \tag{10.2}$$

where T is the temperature of the colder body. By comparing relations (10.1) and (10.2) we come to the conclusion that $\hbar c/r_g$ can be identified with the temperature of the colder body, i.e. of the black hole, in complete agreement with the result obtained by dimensional arguments.

To derive the numerical factor in the relation $T \sim \hbar c / r_g$, we consider the following problem ($T.\ Padmanabhan$, 1999). Let a semiclassical wave packet of a massless field propagate from a point $r_0 = r_g + \varepsilon$ close to the horizon, to a distant point r ($\varepsilon \ll r_g$, $r \gg r_g$). Since the wave packet is semiclassical, its motion can be described by relations obtained in section 6.6 for a point-like particle. According to (6.29), the time of this packet propagation from r_0 to r is

$$t = r - r_0 + r_g \ln \frac{r - r_g}{r_0 - r_g} \approx r + r_g \ln \frac{r}{\varepsilon} \qquad (10.3)$$

(we note that notations r_0 and r are interchanged here as compared to (6.29)). If the frequency at $r_0 = r_g + \varepsilon$ is ω_0, then at $r \gg r_g$ it becomes

$$\omega_1 = \omega_0 \sqrt{g_{00}(r_0 = r_g + \varepsilon)} \approx \omega_0 \sqrt{\frac{\varepsilon}{r_g}} .$$

Since, in virtue of (10.3),

$$\frac{\varepsilon}{r} = \exp \left(-\frac{t - r}{r_g} \right),$$

the frequency ω_1 depends on time as

$$\omega_1 = \omega_0 \sqrt{\frac{r}{r_g}} \exp \left(-\frac{t - r}{2 r_g} \right),$$

and the phase of the wave is

$$\int dt\, \omega_1 = -2 \omega_0 \sqrt{r r_g} \exp \left(-\frac{t - r}{2 r_g} \right).$$

The spectral function of the wave packet at large distances looks as follows:

$$f(\omega) \sim \int dt e^{i \omega t} \exp \left(-2 i \omega_0 \sqrt{r r_g}\, e^{-(t - r)/2 r_g} \right).$$

With the change of variables

$$y = 2 \omega_0 \sqrt{r r_g}\, e^{i \pi / 2} e^{-t / 2 r_g},$$

we express this integral via Γ-function:

$$f(\omega) \sim (2 \omega_0 \sqrt{r r_g})^{2 i \omega r_g} e^{-\pi \omega r_g} \Gamma(-2 i \omega r_g)$$

(factors, independent of ω, are omitted here). In result, the spectral density of the wave packet at large distances is:

$$|f(\omega)|^2 \sim e^{-2\pi\omega r_g}|\Gamma(-2i\omega r_g)|^2$$

$$= \frac{\pi}{\omega r_g}\frac{1}{e^{4\pi\omega r_g}-1} = \frac{\pi}{\omega r_g}\exp(-4\pi\omega r_g) \tag{10.4}$$

(we recall that this is a semiclassical wave packet, so that $\omega r_g \gg 1$). In a remarkable way, the spectral density of a signal, that arrives at infinity from the vicinity of the horizon, is completely universal. And if one goes over in it from the frequency ω to the energy $\hbar\omega$, it gets clear that the leading exponential factor in (10.4) corresponds to the high frequency asymptotics of the Boltzmann distribution with the temperature

$$T = \frac{\hbar c}{4\pi r_g} = \frac{\hbar c^3}{8\pi k M}. \tag{10.5}$$

This expression for the black hole temperature was obtained by *S. Hawking* (1974).

The inevitable result of the finite temperature T of a black hole is the conclusion that in fact it radiates. Black hole produces not only photons and neutrinos with energies on the order of T, but particles of non-vanishing rest mass m as well (only if its temperature is sufficiently high). Thus, one of the most amazing properties of black holes is that they shine!

V.N. Gribov was the first who made this conclusion.[1] One of his arguments was as follows. The uncertainty relation $\Delta E \Delta t \geq \hbar$ allows the creation of pairs of particles from vacuum for the time t that does not exceed $\hbar/E$; here E is the total energy of the pair ($2mc^2$ for massive particles). The gravitational field near the horizon is very strong, so that the energy conservation by itself allows one of the particles to be absorbed by the black hole, and the second one to go to infinity. In quantum mechanics, due to the tunneling effect of such a sort, the processes of particle creation become possible. In particular, the creation of electron-positron pairs in strong electric fields has not only been studied theoretically for a long time, but has been observed experimentally in the heavy ion collisions. In fact, an analogous phenomenon can serve as an explanation of the black hole radiation.

One more of Gribov's arguments is worth mentioning here: a black hole certainly cannot confine radiation with wave length exceeding its gravitational radius. The correspondence is obvious of this argument with the expression (10.5) for the temperature, i.e. for the frequency where the exponential fall down of the intensity starts.

---

[1] *Gribov* precisely formulated the statement that black holes radiate in discussions taking place in 1971 or 1972. This was told to me independently by A.D. Dolgov, D.I. Diakonov, L.B. Okun', who had been present at those discussions. One can only regret that Gribov did not publish this result, apparently he considered it self-evident. In 1974 radiation of black holes was predicted independently by *S. Hawking*.

In fact, for real black holes the temperature (10.5) is negligibly small. In particular, for the mass comparable with that of the Sun it is only about 10^{-7} K. For instance, for the temperature to be sufficient to produce electron-positron pairs, the lightest particles of nonvanishing rest mass, the black hole mass should be smaller than that of the Sun by 17 orders of magnitude, i.e. it should not exceed 10^{17} g. However, a star with such a small mass cannot compress to its gravitational radius, it cannot turn into a black hole. Such light black holes in principle could arise at the most early stages of the Universe evolution when the matter density was very high.

But could such mini-holes survive since those times? Could their age approach the Universe life time $\tau \sim 10^{10}$ years, or 10^{17} s? The obstacle here is the black hole thermal radiation itself. Let us estimate its intensity I by dimensional arguments. To this end it is sufficient to divide T by the characteristic time, which is nothing but r_g/c:

$$ I \sim \frac{cT}{r_g} \sim \frac{m_p^4 c^4}{\hbar M^2}. \tag{10.6} $$

We have introduced here the so-called Planck mass

$$ m_p = \left(\frac{\hbar c}{k} \right)^{1/2} = 2.2 \times 10^{-5} \text{g}. \tag{10.7} $$

On the other hand, obviously, $I = -c^2 dM/dt$. Solving the differential equation

$$ \frac{dM}{dt} = - \frac{m_p^4 c^2}{\hbar M^2}, $$

we find that to survive until our time a black hole should have an initial mass

$$ M > m_p \left(\frac{\tau}{t_p} \right)^{1/3} \sim 10^{15} \, g. \tag{10.8} $$

Here t_p is the so-called Planck time

$$ t_p = \frac{\hbar}{m_p c^2} = \left(\frac{\hbar k}{c^5} \right)^{1/2} = 0.54 \times 10^{-43} \, \text{s}. \tag{10.9} $$

Together with the energy, a black hole also loses its mass. Then, according to relation (10.6), the intensity of its radiation grows, it shines brighter and brighter. The gravitational radius of a black hole gets smaller and smaller. How does this process end? Obviously, a star cannot radiate more energy than it has. The radiation certainly stops when the black hole temperature becomes comparable to its rest energy, at

$$ Mc^2 \sim T \sim \frac{m_p^2 c^2}{M}, $$

i.e. when the mass of such a mini-hole decreases to the Planck mass:

$$M \sim m_p \,.$$

Here our semiclassical consideration of quantum effects in the vicinity of black holes, and in quantum gravity in general, becomes inapplicable. Here a consistent quantum theory of gravity is necessary. However, such a theory does not exist up to now.

Let us go back to the question of whether such bright mini-holes, arising at early stages of the Universe evolution, could survive until our time. Simple estimates with formula (10.8) demonstrate that the mass of the brightest among such relics looks quite modest, it is about 10^{15} g. However, the last stage of its evolution, just before reaching the Planck scale, should be very impressive: an explosion with a power of thousands of the biggest hydrogen bombs. These phenomena have not been observed by astronomers up until now.

It is instructive to look at relation (10.5) from a somewhat different point of view. It demonstrates that the mass of a black hole, and hence its energy as well, decreases as the temperature increases. In other words, the heat capacity of a black hole is negative. This unusual property is in no way special to black holes, it is quite typical for gravitating systems in general.[2] As to a black hole, its negative heat capacity is directly related to the instability caused by radiation. Let us recall, however, the classical instability of the orbit of a bound electron in the Coulomb field. It is also caused by radiation, but is finally stabilized by quantum effects. In the case of black holes as well, it is natural to assume that on the Planck scale their semiclassical radiative instability is stabilized by quantum effects.

In conclusion of this section, we pay attention to the following important fact related to the radiation of black holes. For the typical time interval $\Delta t \sim r_g/c$ between the acts of radiation, the uncertainty of the energy of a black hole is $\Delta E \sim \hbar/\Delta t \sim \hbar c^3/kM$. The corresponding uncertainty in the gravitational radius is ($J.$ $York$, 1983)

$$\Delta r_g \sim \frac{k\Delta M}{c^2} \sim \frac{k\Delta E}{c^4} \sim \frac{\hbar}{Mc} \,.$$

It is only natural to believe that at least due to this uncertainty, the time of the fall of a point-like particle to the horizon, which is logarithmically divergent in the classical approach (see section 6.6), becomes finite:

$$t \simeq r_g \ln \frac{r_g}{\Delta r_g} \simeq r_g \ln \frac{M^2}{m_p^2} \,.$$

And the fact that the arising logarithm is very large, $\ln M^2/m_p^2 \simeq 10^2$, in the present case is not of much importance.

---

[2] See L.D. Landau and E.M. Lifshitz, *Statistical Physics*, part 1, § 21.

Problem

10.1. Prove relation (10.6), starting from the Stefan – Boltzmann law.

10.2 Entropy, Horizon Area, and Irreducible Mass. Holographic Bound. Quantization of Black Holes

Now, when the temperature of a black hole has been found, the calculation of its entropy becomes an elementary problem. The well-known thermodynamical formula

$$dE = TdS \qquad (10.10)$$

relates the increase of the energy E of a body to the increase of its entropy S. In our case T is given by formula (10.5), and $E = Mc^2$. Solving the arising differential equation

$$dM = \frac{\hbar c}{8\pi k M} dS$$

with the natural boundary condition

$$S = 0 \quad \text{for} \quad M = 0,$$

we find

$$S = \frac{4\pi k M^2}{\hbar c}.$$

It is convenient to introduce the so-called Planck length

$$l_p = \left(\frac{\hbar k}{c^3}\right)^{1/2} = 1.6 \times 10^{-33} \text{ cm}. \qquad (10.11)$$

Then we arrive at the following remarkable relation between the entropy of a Schwarzschild black hole and the area of its horizon $A = 4\pi r_g^2$:

$$S = \frac{\pi r_g^2}{l_p^2} = \frac{A}{4l_p^2}. \qquad (10.12)$$

The corresponding analysis for a charged black hole is more intricate. In the Schwarzschild case, the horizon area A depends on the only parameter, the mass M of a black hole. Therefore, the adiabatic invariance of A means that M is also an adiabatic invariant. But what happens with the Reissner – Nordström black hole when a small charge e is lowered adiabatically to its horizon? What remains constant, the horizon area or the mass (if either)?

To answer this question, we resort again to a thought experiment. To simplify the electrostatic part of the problem, we modify the experiment described in the previous section as follows. We will consider a thin spherical shell consisting of small charged particles, each of them attached by its own rope to its own fly-wheel. All the particles are lowered synchronously and adiabatically to the black hole. When the shell reaches the horizon, the charge of the black hole changes from the initial value q to $q + e$, where e is the total charge of the shell. As this was the case in the thought experiment of section 10.1, the rest mass of the shell adds nothing by itself to the mass of the black hole. The latter changes only due to the difference between the final and initial values of the total electrostatic energy (see section 6.7). Since the electric field of a charged shell exists only outside of it, this difference is as follows:

$$\frac{(q+e)^2}{2} \int_{r_1}^{\infty} \frac{dr}{r^2} - \frac{q^2}{2} \int_{r_1}^{\infty} \frac{dr}{r^2} = \frac{(q+e)^2}{2r_1} - \frac{q^2}{2r_1} = \frac{qe}{r_1} + \frac{e^2}{2r_1}. \quad (10.13)$$

Here r_1 is the radius of the new horizon, it has changed as compared to the initial value $r_{rn} = kM + \sqrt{k^2M^2 - kq^2}$, together with the total charge. Thus, to first order in small e the resulting correction to the mass is

$$\Delta M = \frac{eq}{r_{rn}}. \quad (10.14)$$

We have taken into account here that to zeroth order in e $r_1 = r_{rn}$.

When the mass and the charge of a Reissner – Nordström black hole change by ΔM and Δq, respectively, the resulting total change of the horizon area

$$A_{rn} = 4\pi r_{rn}^2 = 4\pi \left(kM + \sqrt{k^2M^2 - kq^2} \right)^2 \quad (10.15)$$

is

$$\Delta A_{rn} = \frac{8\pi r_{rn}k}{\sqrt{k^2M^2 - kq^2}} (\Delta M \, r_{rn} - \Delta q \, q). \quad (10.16)$$

With $\Delta M = eq/r_{rn}$ and $\Delta q = e$, it vanishes for a nonextremal black hole (with $q^2 < kM^2$). Therefore, it is the horizon area of a Reissner – Nordström black hole, but not its mass, which remains constant under the adiabatic change of the charge.

It is useful to introduce the so-called irreducible mass M_{ir} of a black hole (D. Christodoulou, R. Ruffini, 1970, 1971) related to its area A as follows:

$$A = 16\pi k^2 M_{ir}^2.$$

Of course, for a Schwarzschild black hole M_{ir} coincides with M. For a Reissner – Nordström black hole

$$M_{ir} = \frac{1}{2} \left(M + \sqrt{M^2 - q^2/k} \right).$$

Solving this equation for M, we obtain

$$M = M_{ir} + \frac{q^2}{4kM_{ir}}, \quad \text{or} \quad M = M_{ir} + \frac{q^2}{2r_{rn}}. \quad (10.17)$$

The last form of this solution has a simple physical interpretation: the total mass (or total energy) M of a charged black hole consists of its irreducible mass M_{ir} and of the energy $q^2/2r_{rn}$ of its electric field in the outer space $r > r_{rn}$.

Now we will briefly discuss black holes with internal angular momentum. Unfortunately, the solution of the equations that describe these black holes (R. Kerr, 1963) is extremely tedious.[3] Therefore, we resort to some plausible arguments that will allow us to guess the correct result for the area and irreducible mass of a rotating black hole without mentioned tedious calculations.

According to equation (10.17), the difference between the mass and the irreducible mass of a charged black hole is due to the energy of its Coulomb field. It is natural to assume that for a rotating black hole this difference is due to the kinetic energy of its rotation. Then the simplest relation between M and M_{ir}, with the account for the possible relativistic nature of this rotation, is

$$M^2 = M_{ir}^2 + \frac{J^2}{r_k^2} \; ; \tag{10.18}$$

here J is the internal angular momentum of the rotating black hole, r_k is the radius of its horizon. With $r_k = 2kM_{ir}$, (10.18) can be rewritten as

$$M^2 = M_{ir}^2 + \frac{J^2}{4k^2 M_{ir}^2} \; . \tag{10.19}$$

Solving this equation for M_{ir}^2, we obtain

$$2M_{ir}^2 = M^2 + \sqrt{M^2 - J^2/k^2} \; . \tag{10.20}$$

In this way, our guess (10.18) results in the correct formula for the horizon area of a Kerr black hole:

$$A_k = 8\pi \left(k^2 M^2 + \sqrt{k^4 M^2 - k^2 J^2} \right), \tag{10.21}$$

that follows from accurate calculations. Besides, these calculations demonstrate that the horizon surface of a rotating black hole is spherical. And this is also one of the assumptions made in fact in our initial formula (10.18).

We note that, according to formula (10.21), the internal angular momentum J of a Kerr black hole is bounded by the condition $J^2 \le k^2 M^4$. The Kerr black hole with $J^2 = k^2 M^4$ is called extremal.

In this case as well, thought experiments demonstrate that the horizon area of a nonextremal Kerr black hole remains constant under adiabatic change of the internal angular momentum.

---

[3]Even in the book by L.D. Landau and E.M. Lifshitz *The Classical Theory of Fields* (see §104 therein), instead of the corresponding solution of equations of GR, only a footnote is given: "In literature there is no constructive analytical derivation of the Kerr metric, adequate to its physical meaning, and even the direct check of this solution of the Einstein equations demands tedious calculations."

To summarize, the horizon area of nonextremal black holes does not change under considered adiabatic processes. Therefore, in the general case as well, the entropy of a black hole is related to the horizon area by the same formula (10.12) (of course, with the corresponding value of the gravitational radius).

The fact of proportionality between S and A was established by *J. Bekenstein* (1973).

The second law of thermodynamics imposes serious limitations on possible processes not only in the common life. It plays an important role in the physics of black holes as well. In particular, the following statement follows from it: for any interaction among black holes the sum of the areas of their horizons increases or remains constant. Originally this result was obtained by *S. Hawking* (1971), but with quite different arguments.

It is appropriate to mention here the so-called holographic bound (*J. Bekenstein*, 1981; *G. 't Hooft*, 1993; *L. Susskind*, 1995). According to it, the entropy S of any spherical nonrotating body confined inside a sphere of area A is bounded as follows:

$$ S \le \frac{A}{4l_p^2}, \tag{10.22} $$

with the equality attained only for a body that is a black hole.

A simple intuitive argument confirming this bound is as follows. Let us allow the discussed spherical body to collapse into a black hole. Due to the spherical symmetry, this process is not accompanied by radiation or any other loss of matter. Therefore, during the collapse the entropy increases from S to S_{bh}, or at least remains constant. But the resulting horizon area A_{bh} is certainly smaller than the initial confining one A. Then, using relation (10.12) for a black hole, we arrive, through the obvious chain of (in)equalities

$$ S \le S_{bh} = \frac{A_{bh}}{4l_p^2} \le \frac{A}{4l_p^2}, $$

at the discussed bound (10.22).

The result (10.22) can be formulated otherwise. Among the spherical surfaces of a given area, it is the surface of a black hole horizon that has the largest entropy.

The holographic bound looks rather surprising since according to the common experience the entropy of a body is proportional to the volume of this body, but not to the area of its surface. However, usually limit (10.22) is so mild quantitatively that no contradiction with the common experience arises.

We consider now the temperature of charged and rotating black holes. For a rotating black hole the thermodynamic relation (10.10) generalizes to

$$ dM = TdS + \omega dJ, \tag{10.23} $$

where ω plays the role of the rotation frequency. Now, differentiating with respect to S at $J = \text{const}$ the expression for entropy (following directly from (10.12) and (10.21))

$$S = 2\pi \left(kM^2 + \sqrt{k^2 M^4 - J^2} \right), \tag{10.24}$$

we find

$$T = \frac{\partial M}{\partial S} = \frac{\hbar \sqrt{k^2 M^4 - J^2}}{4\pi k M (kM^2 + \sqrt{k^2 M^4 - J^2})}. \tag{10.25}$$

For a charged black hole the analogue of relation (10.23) is

$$dM = TdS + \phi dq, \tag{10.26}$$

where ϕ is the electrostatic potential. Differentiating the expression for entropy in this case

$$S = 2\pi k \left(M + \sqrt{M^2 - q^2/k} \right)^2,$$

we obtain

$$T = \frac{\partial M}{\partial S} = \frac{\hbar \sqrt{M^2 - q^2/k}}{2\pi k (M + \sqrt{M^2 - q^2/k})^2}. \tag{10.27}$$

The important conclusion follows from relations (10.25) and (10.27): the temperature of extremal black holes is equal to zero.

However, the radiation of extremal black holes in no way vanishes, though it certainly is not of a thermal nature. For an extremal charged black hole the nature of this radiation is the particle production by its electric field. This radiation is bounded by the condition $\Delta(kM^2 - q^2) \geq 0$. Obviously, in this case the loss of energy should be accompanied by the loss of charge, i.e. only charged particles can be radiated, all of them having finite rest mass. In the natural situation when e is comparable to the electron charge, the upper limit on the mass μ of a radiated particle looks quite liberal:

$$\mu \leq \sqrt{\alpha} m_p, \tag{10.28}$$

where $\alpha = 1/137$, and m_p is the Planck mass (see (10.7)). Clearly, an extremal black hole (of course, if its mass is sufficient) can radiate any known charged elementary particles, W-boson and t-quark included.

The radiation of a rotating extremal black hole can be explained in an analogous way (*I.B. Khriplovich, R.V. Korkin*, 2002). The loss of a charge by a charged black hole is due in fact to the Coulomb repulsion between this black hole and particles with the same sign of charge. In the present case the reason is the spin-spin interaction: particles (massless mainly) whose total angular momentum is parallel to that of a black hole are repelled from it.

Generally speaking, these mechanisms are operative, in line with the thermal one, for nonextremal black holes as well.

It should be noted that neither the horizon surface, nor, consequently, the entropy of a black hole turn to zero in the extremal case, i.e. for the vanishing temperature. This is in contradiction with the Nernst theorem, or with the so-called third law of thermodynamics, according to which the entropy of a system should vanish when the temperature tends to zero. However, there are no special reasons for anxiety here. In fact, the Nernst theorem is valid only under the condition that the state of a system is nondegenerate at zero temperature. This is the case indeed for stable ground states of common thermodynamic systems. However, due to the mentioned nonthermal radiation, the state of an extremal black hole is in fact metastable one.

We come back now to the adiabatic invariance of the horizon area of a nonextremal black hole. It is well-known that the quantization of an adiabatic invariant is perfectly natural. And just on this argument is based the idea of quantizing the horizon area of black holes proposed by *J. Bekenstein* (1974). Once this hypothesis is accepted, the general structure of the quantization condition for large quantum numbers gets obvious, up to an overall numerical constant β. The quantization condition for the horizon area A should be

$$A = \beta\, l_p^2\, N, \tag{10.29}$$

where N is some large quantum number. Indeed, the presence of the Planck length squared $l_p^2 = k\hbar/c^3$ is only natural in this quantization rule. Then, for the horizon area A to be finite in the classical limit, the power of N here should be the same as that of $\hbar$ in l_p^2. This argument can be checked by considering any expectation value in quantum mechanics, nonvanishing in the classical limit. It is worth mentioning that there are no compelling reasons to believe that N should be an integer. Neither are there compelling reasons to believe that the spectrum (10.29) is equidistant.

However, at present it is not exactly clear how black holes are quantized. We stay, at best, within the semiclassical approximation to the quantum theory of gravity, which has not been built up to now.

Problems

10.2. Find maximum energy liberated under the fusion of two black holes with masses m_1 and m_2.

10.3. Derive relations (10.25) and (10.27).

10.4. Prove condition (10.28).

Index

Adiabatic invariance, 104, 109, 110, 114
Age of the Earth, 101
Age of the Universe, 101

Bianchi identity, 22, 93

Carnot formula, 104
Christoffel symbol, 8, 16, 17, 25, 41–43,
    45, 58, 96
    transformation law, 14, 42
Cosmological term, 27, 28, 46, 101
Critical density, 100
Current density, 5, 6, 30

Dark matter, 100
Doppler effect, 89

Eikonal, 33
    equation, 34
Einstein spaces, 28
Energy-momentum tensor, 6, 27, 29, 31,
    81, 86, 98
    of gravitational field (pseudotensor),
    82
    corresponding to cosmological term,
    28
    of dust, 30
    of electromagnetic field, 43
    of point-like particle, 43
Equation of motion, 5, 8, 9, 24, 29, 30,
    68, 84
Equivalence principle, 6, 7, 15, 73, 81,
    82, 90
Extremal black hole, 58, 111

Four-velocity, 5, 8
Frenkel – Bargman – Michel – Telegdi
    (FBMT) equation, 68
    noncovariant FBMT equation, 71

Gauss theorem, 57
Geodesic, 8, 9, 18–20, 24, 25
Geodesic deviation, 24, 25
Gravitational radius, 9, 54, 55
    of the Earth, 9
    of the Sun, 9
Graviton, 79
    massless, 79
Gyromagnetic ratio, 67
Gyroscope precession, 65

Hamilton-Jacobi equation, 54
Harmonic coordinates (gauge), 6, 29,
    34, 46–48, 51, 52, 77, 78, 80
Helicity, 79
Horizon, 55
Hubble constant, 100

Impact parameter, 34
Integrals of motion, 8, 49
Interval, 5, 7, 8, 12, 44, 46, 47, 51, 55,
    63, 94–96, 99
Irreducible mass, 110
Isotropic coordinates, 47–49, 53

Kerr black hole, 111

Lense – Thirring effect, 66

Locally inertial frame, 7, 9, 15, 21–24, 27, 41, 42, 71, 88, 90
Lorentz gauge, 5, 21
Lorentz transformation, 68
Lorentz transformations, 6, 9, 70, 71, 88

Maxwell equation, 5, 17, 29, 57, 58

Newton constant, 6, 27, 63
Normal coordinates on geodesic, 9, 24

Parsec, 100

Quadrupole moment precession, 71

Radiation
  of black holes, 106
  of extremal black holes, 113
  of ultrarelativistic particles
    angular distribution, 89, 90, 92
    formation length, 89
    frequency spectrum, 89, 92
    total intensity, 88, 91, 92
  quadrupole, 81, 83
Rank, 6, 7, 11, 12, 14, 17, 22, 31
Reissner – Nordtröm solution, 58, 109, 110
Relativistic spin precession in gravitational field, 71, 73
Ricci rotation coefficient, 72–74
Ricci tensor, 22, 24, 28, 44, 96, 97

Riemann tensor
  number of components, 22, 23
  on cone surface, 24
  on torus surface, 24
  symmetries, 21, 22
Rotation
  shift of interference fringes, 65
Runge – Lenz vector, 66

Scalar curvature, 22–24, 41, 93, 94, 96, 97
Schwarzschild solution, 44
Signature, 7
Spin-orbit interaction
  gravitational, 62, 74
  in hydrogen atom, 71
Superposition principle, 5, 30

Thomas precession, 70
Tidal forces, 25, 56
Torsion tensor, 15

Uncertainty relation, 104
  for horizon radius, 108

Variational principle, 2, 7, 8, 16, 43
Volume element, 12

Weak gravitational field, 6, 8, 9, 33, 53, 74